CHEMISTRY RESEARCH AND APPLICATIONS

CHEMICAL VAPOR SYNTHESIS OF INORGANIC NANOPOWDERS

Chemistry Research and Applications

Additional books in this series can be found on Nova's website
under the Series tab.

Additional E-books in this series can be found on Nova's website
under the E-book tab.

Nanotechnology Science and Technology

Additional books in this series can be found on Nova's website
under the Series tab.

Additional E-books in this series can be found on Nova's website
under the E-book tab.

CHEMISTRY RESEARCH AND APPLICATIONS

CHEMICAL VAPOR SYNTHESIS OF INORGANIC NANOPOWDERS

H. Y. SOHN

Nova Science Publishers, Inc.
New York

For permission to use material from this book please contact us:
Telephone 631-231-7269; Fax 631-231-8175
Web Site: http://www.novapublishers.com

LIBRARY OF CONGRESS CATALOGING-IN-PUBLICATION DATA

Chemical vapor synthesis of inorganic nanopowders / editor, H.Y. Sohn.
p. cm.
Includes bibliographical references and index.
ISBN 978-1-62100-002-0 (hardcover)
1. Nanoparticles. 2. Chemical vapor deposition. 3. Organic compounds--Synthesis. I. Sohn, Hong Yong.
TA418.9.N35C47 2011
620.1'18--dc23

2011028038

Published by Nova Science Publishers, Inc. † New York

Dedicated to
Victoria for all her sacrifices,
Berkeley for his inner strengths, and
Edward for his unlimited promises

CONTENTS

PREFACE

Materials in the form of ultrafine or nanosized powders display many useful physical, mechanical and chemical properties. The properties of nano-scaled materials, which are significantly different from atoms and bulks materials, arise mainly from large fraction of surface atoms, high surface energy and reduced imperfections. The effects of grain size on mechanical properties have been recognized and generated much interest. It would therefore be advantageous if compounds can be produced in fine powder form from inexpensive raw materials using less energy. Their small sizes also improve hardness, fracture toughness and low-temperature ductility as well as catalytic and interfacial processes. Nanomaterials also enable lower processing temperatures compared with bulk materials and faster reaction time due to their higher surface reactivity. Thus, the production of nanosized powder is an important industrial process and is explored in this technical and informative book.

Chapter 1

INTRODUCTION

1.1. PROMISES AND APPLICATIONS OF NANOMATERIALS

Materials in the form of ultrafine or nanosized powders display many useful physical, mechanical and chemical properties [Lalena, 2008; Vollath, 2008; Kelsall, 2005; Cao, 2004; Edelstein and Cammarata, 1998]. The properties of nano-scaled materials, which are significantly different from atoms and bulks materials, arise mainly from (1) large fraction of surface atoms; (2) high surface energy; and (3) reduced imperfections [Cao, 2004]. The effects of grain size on mechanical properties have been recognized and generated much interest [Kecskes et al., 2007; Lei et al., 2007]. With large specific surface areas, they are desirable raw materials for making bulk materials by powder processing. In powder metallurgy, powders are consolidated in near net shape and then subjected to secondary processing. It would therefore be advantageous if compounds can be produced in fine powder form from inexpensive raw materials using less energy. Their small sizes also improve hardness, fracture toughness and low-temperature ductility as well as catalytic and interfacial processes [Veprek and Argon, 2001]. Nanomaterials also enable lower processing temperatures compared with bulk materials and faster reaction time due to their higher surface reactivity. Thus, the production of nanosized powder is an important industrial process [German et al., 2006; Wei et al., 2006].

Nanostructured materials also possess other exceptional properties, including light absorption [Nikklason, 1987], magnetism [Okamoto et al., 1987], and superconductivity [Parr and Feder, 1973]. Nanomaterials of metallic elements are regularly alloyed in order to create nanoalloys with much superior thermal and electrical properties. An example is Ag-Al nanoalloys, which have a high potential for electronic applications as a substitute for more expensive Ag nanopowder for use as interconnects. They are also excellent candidates for solar cell application. Nanomaterials have a great potential in microelectronics field based on their small sizes, which contributes to the miniaturization of the devices. Nanoelectronic devices based on new nanomaterials systems and new device structures will contribute to the development of next generation microelectronics. For example, single electron transistor [Postma et al., 2001] and field effect transistor [Keren et al., 2003; Tans et al., 1998; Martel et al., 1998] based on single wall carbon nanotubes are already on the way. Prototype simple logic circuits of carbon nanotubes have already been demonstrated [Bachtold et al., 2001; Collins et al., 2001; Derycke et al., 2001; Rueckes et al., 2000].

Another promising application of the nanomaterials is related to the use of nanofluid to enhance the thermal transport. Nanofluids refer to liquids with nanopowder suspended in them. The presence of small amounts of the suspended nanoparticles greatly enhances the thermal conductivity of the liquid [Patel et al., 2003; Xie et al., 2002a; Xie et al., 2002b; Eastman et al., 2001; Choi et al, 2001].

1.2. Methods of Production

There are numerous methods to produce the nanosized powders [Manikama et al., 2011]. In principle, those methods can be divided into two categories: "bottom-up" methods and "top-down" methods. The "bottom-up" methods involve synthesis of nanoparticles by condensation of atoms or molecular entities in a gas phase or in solution, which includes such methods as sol-gel and other colloidal chemistry methods, aerosol and spray processes, and gas phase reactions. The "top-down" methods involve preparation of the nanostructure from coarser forms, such as milling. Most widely used methods for synthesizing inorganic nanoparticles are described below:

High energy milling is a mechanical method of generating nanopowders. The process involves the mechanical attrition of larger particles down to nanometer size in a mill, typically a ball mill. High energy milling can also be used for mechanical alloying to synthesize composite powders. In this process, cold welding and fracturing of powders are repeated, which results in forming an alloy of fine size. Many ceramic powders [Xin et al., 2005; Miclea et al., 2004; Kong et al., 2001] and alloys as well as composites [Liu et al., 2004; Wagner et al., 1996] have been produced by this method. High energy milling offers the advantages of producing bulk quantities of nanopowder at room temperature. Also, high energy milling can be applied to almost any material and the simplicity of the process also translates to easy industrial scale up. Contamination problems have plagued this technique, but the application of tungsten carbide components and the introduction of inert atmosphere and high vacuum milling system have decreased impurities to acceptable industrial levels. Nevertheless, problems such as low surface area, non-uniform size distributions, and partially amorphous state of the as-prepared powders persist.

Inert gas condensation involves vaporizing the solid into a background gas then mixing the vapor with a cold gas to lower the temperature [Siegel, 1994; Gleiter, 1989]. This method is well suited for production of metal nanoparticles, since many metals vaporize at reasonable rates at attainable temperatures. By including a reactive gas such as oxygen in the cold gas stream, oxides or other compounds of the vaporized materials can also be prepared.

Laser ablation uses a laser to vaporize solid materials in an inert flow reactor [Becker et al., 1998]. This method is good for precursors that have high volatilization temperatures and low vapor pressures such as refractory oxides. The main drawback to this method is that the amount of nanopowder produced is relatively small.

Spark discharge generation involves using electrodes made of the metal to be vaporized in the presence of an inert background gas until the breakdown voltage is reached. The arc formed across the electrodes vaporizes a small amount of metal. This produces very small amounts of nanoparticle with relatively reproducible results.

Ion sputtering involves the use of high velocity ions of an inert gas to vaporize the solid. The advantage of this method is that the composition of the sputtered material is the same as that of the target. The disadvantage of this method is that low pressures are required, which makes further processing of the nanoparticles in aerosol form difficult.

Spray conversion processing involves coordinated steps: Preparation and mixing of an appropriate starting solution, spray drying to form a chemically homogeneous precursor powder, and conversion of the precursor into the desired nanophase powder. This process has been applied for the production of nanosized composite powders [Ban and Shaw, 2002; McCandlish et al., 1992]. It is a scalable technology, which provides the means for producing bulk quantities of nanophase powders at low manufacturing costs. Variations include spray pyrolysis [Messing et al., 1993] and plasma spray [Berndt and Lavernia, 1998].

The flame synthesis method utilizes an open flame for particle synthesis. The temperature of the flame gives the necessary heat for the reactions to occur. This is the most successful method of producing nanoparticles industrially. It is primarily useful for making oxides, since the flame environment is usually oxidizing [Calcote and Keil 1997; Axelbaum 1997; Pratsinis 1997; Zachariah 1994]. The drawback to this method is the high agglomeration of the synthesized particles. Sol-gel processing is a wet chemical synthesis approach that can be used to generate nanoparticles by gelation, precipitation, and hydrothermal treatment [Kung and Ko, 1996]. Size distribution of semiconductor, metal and metal oxide nanoparticles can be manipulated by either dopant introduction [Kyprianidou-Leodidou et al., 1994] or heat treatment [Wang and Ying, 1999]. In hydrodynamic cavitation, nanoparticles are generated through creation and release of gas bubbles inside the sol-gel solution [Sunstrom et al., 1996]. By rapidly pressurizing in a supercritical drying chamber and exposing to cavitational disturbance and high temperature heating, the sol-gel solution is mixed. The erupted hydrodynamic bubbles are responsible for nucleation, growth, and quenching of the nanoparticles. Particle size can be controlled by adjusting the pressure and the solution retention time in the cavitation chamber.

Microemulsions have been used for synthesis of metallic [Kishida et al., 1996], semiconductor [Pileni et al. 1992; Kortan et al., 1990], silica [Arriagada and Osseo-Asare, 1999], barium sulfate [Hopwood and Mann, 1997], magnetic and superconductor [Li et al, 2000; Pillai et al. 1995] nanoparticles.

Gas phase reaction offers the advantages of synthesizing high purity nanopowders with good control over size, shape, and crystal structure as well as easy control of reaction rate. Nanopowders synthesized by this method exhibit excellent properties for consolidation into bulk materials with improved mechanical properties such as material strength, hardness, and toughness and lower sintering temperatures during the compaction process. Gas phase synthesis to prepare nanopowders is conducted by creating a supersaturated vapor of a precursor which is thermodynamically unstable and thus form a condensed phase rather than remain in the gas phase. There are several methods for preparing nanopowders by gas phase synthesis [Swihart, 2003; Kruis and Fissan, 1999; Chang et al., 1994], including the chemical vapor synthesis which is the subject of this monograph. The major difference in such methods lies in the energy source used to achieve a state of supersaturation and the nature of the precursors used.

1.3. Chemical Vapor Synthesis

Chemical vapor synthesis involves introducing vapor phase precursors into a heated reactor and allowing vapor phase nucleation to occur in the reactor rather than depositing the product as a thin film [Sohn et al., 2007]. The precursors can be solid or liquid in nature but are introduced into the reactor as a vapor by sublimation or evaporation. Metal chlorides are generally preferred as the precursors owing to their low volatilizing temperature. Chemical vapor synthesis has been applied to the synthesis of metallic [Ryu et al., 2010a; Choi et al., 2009a; Ryu et al., 2009a; Sohn et al., 2007] and intermetallic and alloy powders [Sohn et al., 2004; Sohn and PalDey, 1998a; 1998b; 1998c; 1998d], carbides [Sohn and Ryu, 2010; Ryu et al., 2008a; 2008b; 2009b; Sohn et al., 2007] and their composites [Ryu et al., 2010a; 2009c; 2008c; 2008d; Sohn et al., 2007], and oxide nanopowders [Ryu et al., 2010b].

Chemical vapor synthesis assisted by thermal plasma involves the use of a plasma flame as the heat source. Plasma flames can facilitate vapor phase reactions by providing sufficient energy for vaporizing precursors and subsequent chemical reactions. The temperature of plasma flame generated is high enough to decompose even reactants of high vaporization temperatures into atoms and radicals, which can then react and condense to form nanosized particles when cooled by mixing with cool gas or expansion through a nozzle.

This book describes the production of inorganic nanopowders by the chemical vapor synthesis (CVS) process in which fine solid particles are made by vapor-phase chemical reactions of precursors. The outstanding features of the CVS process are product particle sizes ranging from 5-200 nm, unique capability to produce uniformly mixed powders of different solids, and ease of dopant addition in one-step synthesis. In the CVS process, the final particle size and morphology of the product are determined by the reaction temperature, residence time, position of reactant feeders, and aerodynamic conditions inside the reactor.

Chapter 2

PRINCIPLES OF CHEMICAL VAPOR SYNTHESIS

ABSTRACT

Chemical vapor synthesis involves complex reaction steps, each of which requiring understanding of many different scientific principles. Out of these, some of the main basic principles involved in the theory and practice of the chemical vapor synthesis (CVS) processes are discussed in this chapter.

2.1. THERMOCHEMICAL CONSIDERATIONS

Before starting any investigation of a chemical process, it is useful to carry out a theoretical thermodynamic calculation to determine the conditions to produce the desired phases. An example of such a calculation is illustrated here using the preparation of aluminum nanopowder by the magnesium reduction of aluminum chloride vapor:

$$AlCl_3(g) + 1.5Mg(g) = Al + 1.5MgCl_2(g) \quad (2\text{-}1)$$

First, a simple calculation of $\Delta G°$ qualitatively indicates the likelihood of the reaction occurring as written at the selected temperature. For the above reaction, $\Delta G°$ is – 29.2 kcal at 1000°C, indicating that the reaction is likely to proceed to the right under most conditions.

More definitive prediction of thermodynamic feasibility of a reaction must include the possibility of forming other species. A number of computer software programs such as HSC Chemistry [Outotec, 2002], which are based on the principle of the Gibbs free energy minimization, are available to calculate the thermodynamic aspects of reactions, as was done by Choi et al. [2010]. The species considered in the equilibrium calculation were the gaseous species: Ar, $MgCl_2$, AlCl, Mg, Mg_2Cl_4, MgCl, $AlCl_3$, $AlCl_2$, MgCl, and Al plus the condensed phases: $MgCl_2$, Al, $AlCl_3$, and Mg. Apart from these species, other species such as Al_2Cl_6 and Al_2Cl_4 were also considered in the preliminary calculations, but they were found to be present in negligible amounts in the ranges of conditions examined in this work.

Figure 2.1 shows the equilibrium composition as a function of temperature. The synthesis of aluminum with the precursors is seen to be feasible over a wide temperature range. $MgCl_2$ is present as a condensed phase below about 1100°C because of its high boiling point,

1412ºC. As the temperature increases to 1200ºC, the amounts of AlCl(g) and Mg(g) increase while that of Al decreases. Although not shown in this figure, Al_3Mg_2 and $Al_{12}Mg_{17}$ are known to occur below 450ºC in the phase diagram of the Al-Mg system [ASM International, 1992]. To prevent the formation of the intermetallic phases, the product must be quenched rapidly.

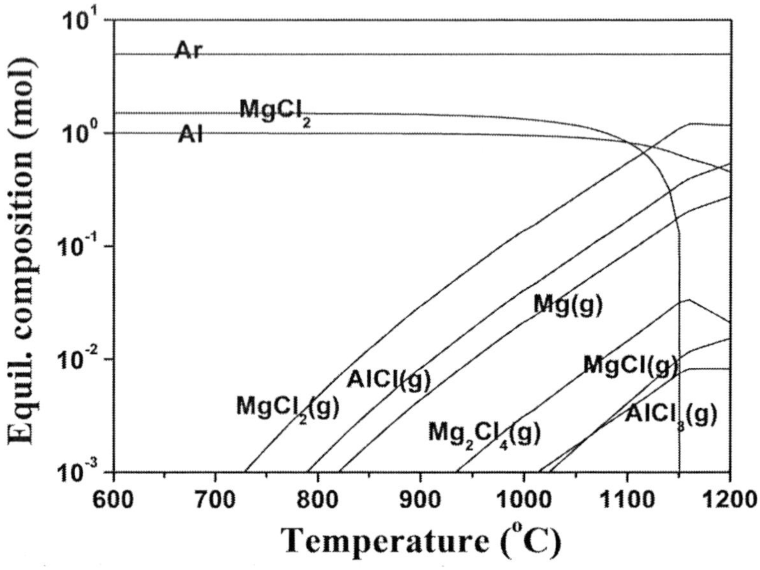

Figure 2.1. Equilibrium composition vs. temperature for a mixture of 1 mol $AlCl_3$-1.5 mol Mg-5 mol Ar under 101.3 kPa pressure. [Adapted from Choi et al. (2010)].

It is noted that the use of such software requires the inclusion of a sufficient number of possible gaseous species and judicious selection of condensed phases. Often, especially for a complex reaction system, an iterative procedure is needed that requires the removal of condensed phases appearing in small quantities until the phase rule is satisfied.

A somewhat more complex system is exemplified by the synthesis of intermetallic compounds such as titanium aluminides, which involves the vapor-phase reduction of mixed metal chlorides by magnesium. Sohn and PalDey [1998a] carried out equilibrium calculations for the Ti-Al-Mg-Cl-Ar system using HSC Chemistry [Outotec, 2002] software. According to the Gibbs phase rule, in a five-component system, when total pressure, temperature, argon partial pressure, and Mg/Ti are fixed, two condensed phases and a gas phase should coexist at equilibrium. The pairs of intermetallic phases that can coexist at equilibrium can be found in the Ti-Al binary phase diagram. Those pairs of condensed phases were tested one at a time in the calculation until the stable phases were determined. The melting point of $MgCl_2$ formed is 708°C, and its equilibrium vapor pressure at 900°C is 0.01 atm. Argon was used in large excess to keep the partial pressure ratio Mg/Ar low enough to ensure that all $MgCl_2$ be present in the gas phase. At higher values, the partial pressure of argon does not materially affect the equilibrium results.

The results of the equilibrium calculation are presented in Figures 2-2 through 2-5. It is observed that the input molar ratios $AlCl_3/TiCl_3$ and $Mg/TiCl_3$ and reaction temperature have significant effects on the equilibrium contents. Figure 2.2(a) shows the effect of the input molar ratio $AlCl_3/TiCl_3$ on the equilibrium contents at 900°C. The input ratio of $Mg/TiCl_3$ is 10. As the

$AlCl_3/TiCl_3$ ratio increases starting from 0.1, the formation of Ti_3Al increases at the expense of metallic Ti and reaches a maximum of 99 mol pct at the $AlCl_3/TiCl_3$ ratio of 0.3. Beyond that, TiAl becomes favorable compared with Ti_3Al. The maximum amount of TiAl is formed at the $AlCl_3/TiCl_3$ ratio of 1. Beyond this ratio, the content of TiAl decreases and $TiAl_3$ starts forming. At the ratio of 3, the content of $TiAl_3$ reaches a maximum value of 99 mol pct and then it decreases in favor of metallic Al. At the ratio of 5.5, the mole percent of $TiAl_3$ goes up again, whereas that of Al goes down. It was found that beyond this point, the amount of $TiAl_3$ remains constant, while that of Al decreases continuously with increasing $AlCl_3/TiCl_3$ ratio. In this region, $AlCl_3$ is reduced to AlCl, rather than to Al, as can be seen in the corresponding gas-phase composition presented in Figure 2.2(b). Another set of calculations on the effect of the input molar ratio $AlCl_3/TiCl_3$ on the equilibrium contents was carried out at the input ratio of $Mg/TiCl_3$ = 5.0, as shown in Figure 2.3. The result is similar to the results presented in Figure 2.2(a), except that in this case, Al is not formed because the amount of Mg vapor available to reduce the excess $AlCl_3$ is insufficient.

The effect of the input $Mg/TiCl_3$ ratio on the solid phase contents at the $AlCl_3/TiCl_3$ ratio of 3 is presented in Figure 2.4. The solid phase contained Ti_3Al, TiAl, or $TiAl_3$. Titanium and aluminum were absent at the $AlCl_3/TiCl_3$ ratio of 3. Starting from the $Mg/TiCl_3$ ratio of 0.1, the content of Ti_3Al decreases and that of TiAl increases until it reaches a maximum value of 99 mol pct at the $Mg/TiCl_3$ ratio of 4; beyond this ratio, the content of TiAl decreases and that of $TiAl_3$ increases until it levels off at the ratio of about 10. Metallic Al does not form in this case.

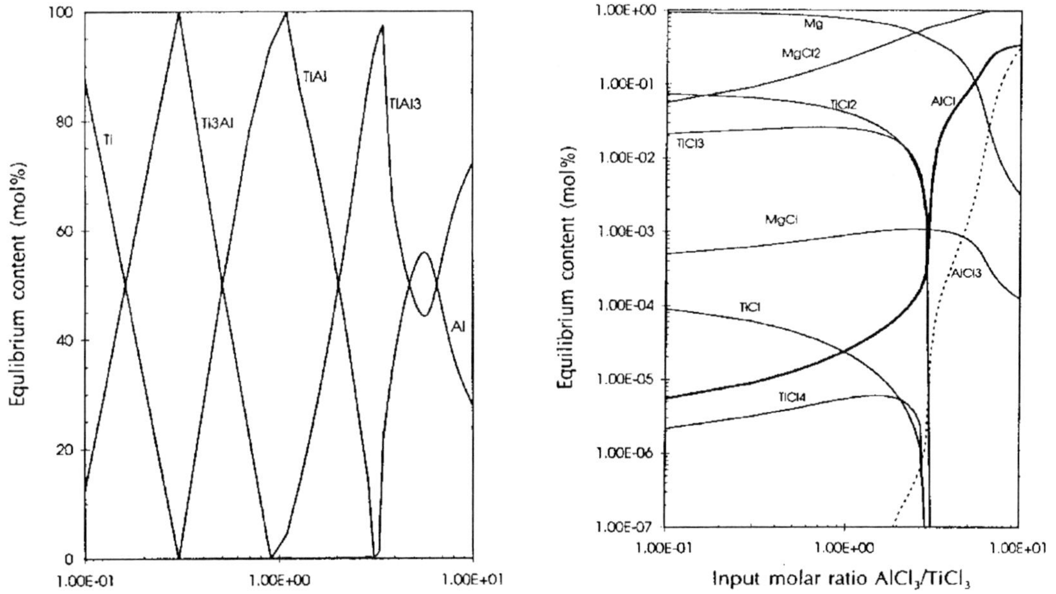

Figure 2.2. (*a*) Calculated effect of the input molar ratio $AlCl_3/TiCl_3$ on the equilibrium solid phase contents in the Ti-Al-Mg-Cl-Ar system (input molar ratio $Mg/TiCl_3$ = 10, total pressure = 101.3 kPa, and temperature = 900°C). (*b*) Calculated effect of the input molar ratio $AlCl_3/TiCl_3$ on the equilibrium gas phase contents in the Ti-Al-Mg-Cl-Ar system (input molar ratio $Mg/TiCl_3$ = 10, total pressure = 101.3 kPa, and temperature = 900°C). [Adapted from Sohn and PalDey (1998a)].

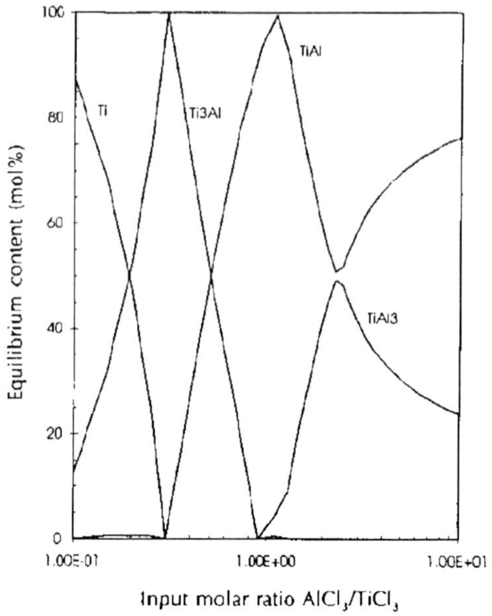

Figure 2.3. Calculated effect of the input molar ratio $AlCl_3/TiCl_3$ on the equilibrium solid phase contents in the Ti-Al-Mg-Cl-Ar system (input molar ratio $Mg/TiCl_3 = 5$, total pressure = 101.3 kPa, and temperature = 900°C). [Adapted from Sohn and PalDey (1998a)].

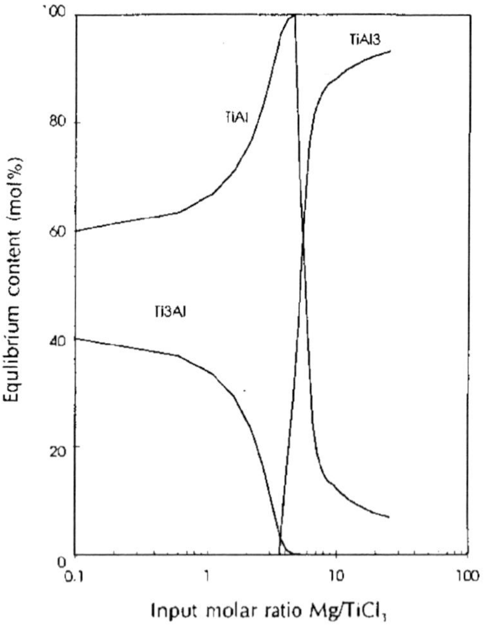

Figure 2.4. Calculated effect of the input molar ratio $Mg/TiCl_3$ on the equilibrium solid phase contents in the Ti-Al-Mg-Cl-Ar system (input molar ratio $AlCl_3/TiCl_3 = 3.0$, total pressure = 101.3 kPa, and temperature = 900°C). [Adapted from Sohn and PalDey (1998a)].

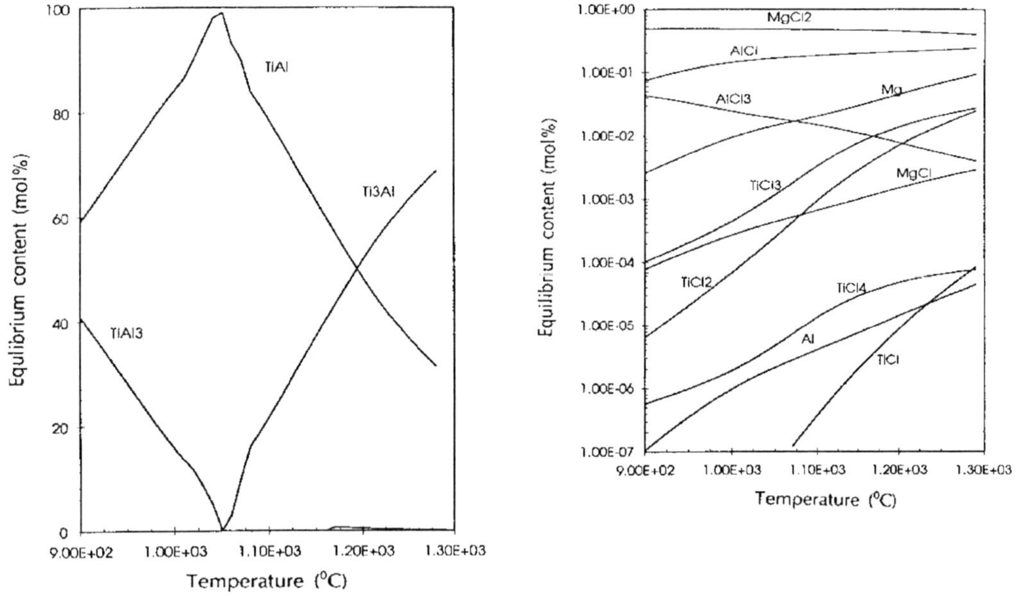

(a) (b)

Figure 2.5. (*a*) Calculated effect of temperature on the equilibrium solid phase contents in the Ti-Al-Mg-Cl-Ar system (input molar ratio $AlCl_3/TiCl_3 = 3.0$, input molar ratio $Mg/TiCl_3 = 5.0$, and total pressure = 101.3 kPa). (*b*) Calculated effect of temperature on the equilibrium gas phase contents in the Ti-Al-Mg-Cl-Ar system (input molar ratio $AlCl_3/TiCl_3 = 3.0$, input molar ratio $Mg/TiCl_3 = 5.0$, and total pressure = 101.3 kPa). [Adapted from Sohn and PalDey (1998a)].

Figure 2.5(a) shows the effect of reaction temperature, in the range 900°C to 1300°C, on the equilibrium contents of the solid phases. The input molar ratios of $Mg/TiCl_3$ and $AlCl_3/TiCl_3$ were at 5.0 and 3.0, respectively. It can be seen that lower temperatures favor the formation of $TiAl_3$, whereas Ti_3Al is favored at higher temperatures. The content of TiAl goes through a maximum at 1050°C. The composition of gas phase as a function of temperature is presented in Figure 2.5(b).

To summarize, the theoretical equilibrium analysis of the Ti-Al-Mg-Cl-Ar system showed the formation of several intermetallic phases along with some pure metals. A single intermetallic phase was formed only at a very narrow range of the partial pressure ratio of reactants; for all other conditions, the solid product was a mixture of two phases. The contents of Ti_3Al and TiAl reached 100 mol pct at $AlCl_3/TiCl_3$ ratios of 0.3 and 1.0, respectively, when $Mg/TiCl_3$ was 10 and the temperature was 900°C.

These types of equilibrium calculations are useful in developing a process, designing experiments and determining optimum conditions to prepare desired materials.

2.2. Chloride-Based Reactions

Metal chlorides are often used as the precursors owing to their low volatilization temperatures. Here, we will summarize developments in the chemical vapor synthesis of nanosized powders by the hydrogen reduction of metal chlorides, which can in general be written as follows:

$$MCl_n(g) + 0.5nH_2(g) = M(s) + nHCl(g) \quad (2\text{-}2)$$

where M represents the metal. The hydrogen reduction of single chlorides for the preparation of metallic ultrafine particles (UFP) has been studied for tungsten and molybdenum [Lamprey and Ripley, 1962], cobalt [Saeki et al., 1978], and nickel, cobalt, and iron [Otsuka et al., 1984]. By reducing vaporized $FeCl_2$, $CoCl_2$, and $NiCl_2$ by hydrogen at 1200 to 1300 K, Otsuka et al. [1984] prepared corresponding metal particles in the size range 52 to 140 nm with up to 99.7 % metal chloride conversion. The synthesis of metal carbide UFP has been practiced by the vapor-phase reduction by hydrogen and hydrocarbon [Zhao et el., 1990; Hojo et al., 1978]. Hojo et al. [1978] produced the UFP of tungsten carbide (WC, W_2C) of 40 - 110 nm size by vapor-phase reaction of the WCl_6–CH_4–H_2 system at 1000 - 1400°C.

Magnesium is a much stronger reducing agent for chlorides than hydrogen. Thus, the reduction of titanium and aluminum chlorides by magnesium is feasible at temperatures around 1000°C, whereas the reduction of these chlorides by hydrogen is not feasible up to nearly 3000°C. Titanium sponge is prepared by the chlorination of rutile, followed by the reduction of the resulting titanium chloride by liquid or gaseous magnesium [Barksdale, 1966]. A reaction between a metal chloride and magnesium vapor can be written as follows:

$$MCl_n(g) + 0.5nMg(g) = M(s) + 0.5nMgCl_2(l,g) \quad (2\text{-}3)$$

where M represents the metal.

Sohn and coworkers [Sohn et al., 2004; Sohn and PalDey, 1998a; 1998b; 1998c; 1998d] applied the basic concepts of the above chloride reduction methods to the chemical vapor synthesis of intermetallic and metal alloy powders. These reactions can in general be written as follows, when hydrogen is used:

$$mMCl_x(g) + nNCl_y(g) + 0.5(mx + ny)H_2 = M_mN_n(s) + (mn + ny)HCl(g) \quad (2\text{-}4)$$

where M and N represent two different metals, x and y being the valences, and M_mN_n the intermetallic compound formed. Sohn and PalDey [1998c] synthesized fine powder (100-200 nm) of Ni_4Mo at 900 to 1100°C using hydrogen as the reducing gas. Sohn and PalDey [1998d] also synthesized nickel aluminide (Ni_3Al) particles (50-100 nm) at 900 to 1150°C using hydrogen as the reducing agent. The fact that aluminum chloride is reduced by this reaction scheme is very significant, because the reduction of $AlCl_3$ alone by hydrogen is thermodynamically unfavorable at these temperatures. The negative free energy of formation of the intermetallic compound makes the overall reaction feasible. Using the same chemical vapor synthesis process, Sohn et al. [2004] prepared ultrafine particles of Fe-CO alloys by the hydrogen reduction of $FeCl_2$-$CoCl_2$ mixtures.

Sohn and PalDey [1998a; 1998b] also synthesized ultrafine powders of the aluminides of titanium and nickel using magnesium as the reducing agent.

2.3. Plasma-Assisted Reactions

Thermal plasma provides many advantages on the synthesis of nanosized powder such as a high processing temperature to vaporize all the reactants completely and rapidly, a high quench rate to form fine powders, and the versatility of a wider range of reactants to choose from. Numerous reports [Lee et al. 2007; Chazelas et al. 2007; Tong and Reddy 2005; Swihart 2003; Fukumasa and Fujiwara 2003; Moriyoshi et al. 1997] on the thermal plasma synthesis of metals, ceramics and composites in recent years have shown that thermal plasma synthesis has a high potential for many industrial applications as one of the most promising methods for producing nanosized powders.

Chemical vapor synthesis assisted by thermal plasma involves the use of a plasma flame as the heat source [Sohn et al., 2007; Ryu et al., 2008a; 2009b; 2009c; 2010a; 2010b]. Plasma flames can facilitate vapor phase reactions by providing sufficient energy for vaporizing precursors and subsequent chemical reactions. The temperature of plasma flame is high enough to decompose the reactants into atoms and radicals, which can then react and condense to form nanosized particles when cooled by mixing with a cool gas or expansion through a nozzle.

The synthesis of tungsten-cobalt nanocomposite powder from a mixture of ammonium paratungstate and cobalt oxide in a plasma reactor is an example [Ryu et al., 2010a]. It is known that APT is easily decomposed to WO_3 by heating [Fait et al. 2008; Madarász et al. 2004; Medeiros et al. 2001; van Put 1995]. The equilibrium composition diagram for the WO_3-Co_3O_4-H_2 mixture was obtained [Ryu et al. 2010a] by the use of HSC Chemistry software developed by Outotec [2002], as shown in Figure 2.6. The diagram shows that it is thermodynamically feasible to synthesize W and Co from the WO_3-Co_3O_4-H_2 mixture in the plasma process. WO_3 was produced in the plasma process when APT powder was injected into the plasma flame with Ar containing no reactant gases, as shown in Figure 2.7(a). The product obtained by hydrogen reduction of a mixture of APT and Co_3O_4 was W-Co composite powder, as shown in Figure 2.7(b).

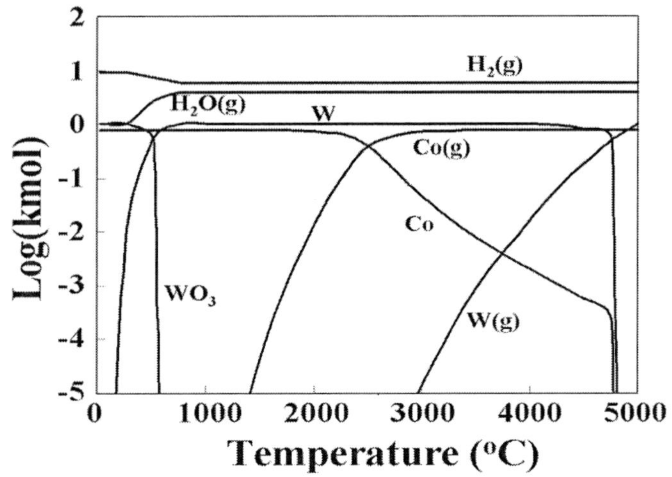

Figure 2.6. Calculated equilibrium composition of product from a mixture of WO_3, Co_3O_4, and H_2 in the amounts of 1, 0.25, and 10 kmol, respectively. [Adapted from Ryu et al. (2010a)].

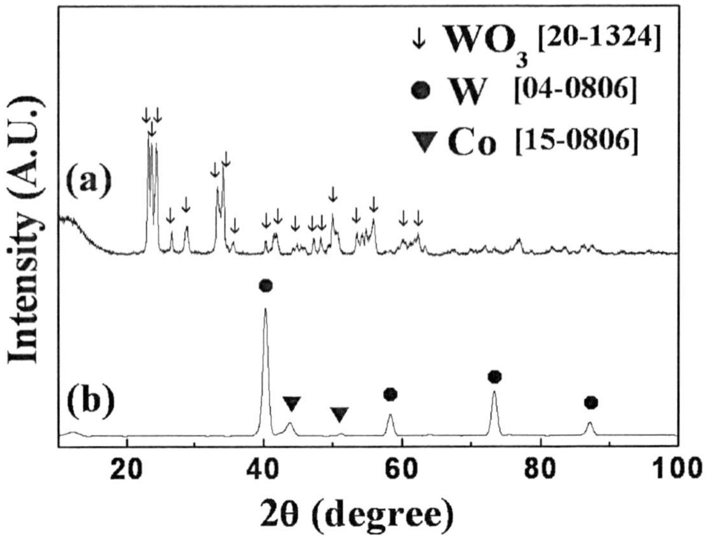

Figure 2.7. X-ray diffraction patterns of the product obtained by the thermal plasma process from (a) APT-Ar and (b) APT-Co_3O_4-H_2 mixture. [The numbers after the phases are the JCPDS file numbers.] [Adapted from Ryu et al. (2010a)].

Crystalline phases were identified by comparing the experimental data with the Joint Committee of Powder Diffraction Standards (JCPDS) files from the International Center for Diffraction Data (ICDD).

The results indicate that W and Co particles were uniformly mixed in the product, which presents an example of the significant feature of the vapor phase reaction to produce very uniformly mixed powders.

2.4. Flame Synthesis

Flame synthesis processes can be divided into the flame vapor synthesis (FVS) and the flame spray pyrolysis (FSP).

In FVS, the precursor is introduced into the flame as a vapor. The important steps in FVS are chemical transformation from precursor to oxide, the nucleation of oxide particles, the collisions of oxide particles, the condensation of oxide vapor onto oxide particles, and the sintering of oxide particles [Friedlander 2000]. However, FVS has the disadvantages of requiring vapor-phase precursors. Also, it is not easy to control the concentration of each precursor during the synthesis of multi-component particles.

In FSP, the precursors are fed into the flame reactor as sprayed droplets. Important advantages of FSP over FVS is that it can easily synthesize multi-component particles and use wide ranges of precursors regardless of their volatility [Chang et al., 2006; Jang et al., 2004; Mueller et al., 2004; Sahm et al., 2004; Strobel et al., 2004; Madler and Pratsinis, 2002; Madler et al., 2002; Tani et al., 2002; Pratsinis, 1998; Bickmore et al., 1998; Bickmore et al., 1996]. In terms of particle formation mechanism, FSP is more complicated than FVS because the evaporation of the droplets has to be considered in addition to the process steps in FVS.

Flame synthesis has been widely used in industry for producing silica particles. Silica nanoparticles are commercially produced by the flame synthesis method because it can provide high production rates at relative low cost.

It allows a good control of the particle size and crystallinity, and is a continuous process without the need for further treatments like calcination [Pratsinis, 1998]. It is considered as a novel technique of producing fine, pure and single-phase particles in the as-prepared state [Madler et al., 2002; Jang, 1999].

Many precursors have been used to form silica particles by FSP, including hexamethyldisiloxane (HMDSO) [Madler et al., 2002], tetraethylorthosilicate (TEOS) [Jang et al., 2006], tetramethylsilane (TMS), tetramethylorthosilicate (TMOS) [Berggren et al., 2005; Kawakami et al., 2000; Dabadie et al., 1995] and tetrapropylorthosilicate (TPOS) [Jossen et al., 2005].

Chapter 3

METALLIC NANOPOWDERS

ABSTRACT

This chapter deals with several examples of preparing metallic nanopowders of wide interest by different chemical vapor synthesis (CVS) techniques. Various methods for characterizing produced nanopowders are also discussed.

3.1. EARLY WORK ON METAL POWDERS

Lamprey and Ripley [1962] were apparently the first investigators that synthesized nanosized (ultrafine) metal powders by the hydrogen reduction of metal chlorides. They produced 20-30 nm size powders of tungsten and molybdenum by reducing, respectively, tungsten chloride and molybdenum chloride vapors with hydrogen at 430-905°C in a quartz-tube reactor. These are chlorides that are less stable than HCl and thus thermodynamically amenable to hydrogen reduction in this temperature range, as shown below:

$$1/6\ WCl_6(g) + 1/2\ H_2(g) = 1/6\ W(s) + HCl(g); \qquad (3\text{-}1)$$

$\Delta G^o = -\ 15.3$ kcal at 1000 K

$$1/5\ MoCl_5(g) + 1/2\ H_2(g) = 1/5\ Mo(s) + HCl(g); \qquad (3\text{-}2)$$

$\Delta G^o = -\ 11.4$ kcal at 1000 K

More definitive determination of feasibility must, however, include the several sub-chlorides that these metal chlorides have. All the sub-chlorides of tungsten and molybdenum are thermodynamically amenable to hydrogen reaction in the above temperature range.

Subsequent related work was carried out by Saeki et al. [1978] who prepared ~10 nm size cobalt powder at 905°C using a similar method and by Otsuka et al. [1984] who prepared 30-110 nm size particles of nickel, cobalt and iron by reducing vaporized $FeCl_2$, $CoCl_2$, and $NiCl_2$ by hydrogen at 750 to 950°C. Ishikawa et al. [1990] reduced $AgCl$, Cu_3Cl_3 and WCl_6 gases by hydrogen at 700 - 1100°C to obtain, respectively, Ag, Cu and W particles of 20-1000 nm diameters. Silver and copper particles were quite spherical, whereas tungsten particles were polyhedral.

The above investigations were done largely by directly contacting the metal chloride vapors carried by an inert gas with hydrogen gas. An exception was the experiments of Lamprey and Ripley [1962] in which a sheath gas flow shrouded the metal chloride-argon stream at the tip of the injector before meeting the hydrogen stream. The sheath gas consisted of argon or chlorine and its purpose was to prevent formation of metal at the gas injection tip. Another variation of the hydrogen reduction of metal hydrides was the use of a hydrogen-chlorine flame as a source of a hot gas mixture that still contained enough hydrogen for the reduction of metal chloride [Zhao et al., 1990]. This process would be feasible only for those metal chlorides that are much less stable than HCl based on one g-atom of chlorine to avoid the effect on equilibrium of the excess HCl generated by the hydrogen-chlorine flame.

3.2. Tungsten Nanopowder – Plasma-Assisted

Tungsten is widely used in high temperature applications such as filaments in light bulbs, cathodes in high power lamps, and rocket nozzles in space crafts as well as in the manufacture of superalloys and catalysts for hydrodesulfurization and hydrodenitrogenation [Bless et al., 2006; Gao et al., 2006; Selcuk and Wood, 2005]. It also finds considerable applications in the form of tungsten carbide as a refractory material based on its outstanding hardness and wear resistance [Sohn and Ryu, 2010; Fang et al., 2009; Ryu et al., 2008a; 2008c; Gao and Kear, 1995].

Tungsten in the form of nanosized powder has the potential to produce very hard, strong, and wear resistant materials when consolidated into compacts. A small particle size lowers the sintering temperature, densifies quickly and yields fine-grained structure when sintered compared with the conventional microcrystalline powders [Ryu et al., 2009d; Wang et al., 2008a; Malewar et al., 2007; German et al., 2006; Wei et al., 2006; Rabe and Wäsche, 1995].

Main raw materials for tungsten are wolframite and scheelite minerals, which are treated to first produce ammonium paratungstate (APT) as an intermediate product. Tungsten oxides are then obtained from APT by calcination between 500°C and 850°C, which is reduced with H_2 to produce the metal [Yih and Wang, 1979]. Nanosized tungsten powder can be produced by various methods such as electrodeposition, sputtering, ball milling, and complicated chemical methods [Gostishchev et al., 2006; Singh Gaur, 2006; Lassner and Schubert, 1999; Welham 1999], but these methods involve many difficult steps. Conventional evaporation and condensation methods cannot be applied to the production of nanosized tungsten powder due to its low vapor pressure [Magnusson et al., 2000]. Although the preparation of tungsten nanopowder by the hydrogen reduction of tungsten chloride in the vapor phase has been investigated, as discussed in the previous section, this requires conversion of tungsten or tungsten oxide to the chloride.

Plasma Synthesis

Recently, a plasma processing technique has been developed for the preparation of nanosized tungsten powder [Ryu et al., 2009a; 2009b], using ammonium paratungstate (APT) (99.9%) as the precursor and reducing its decomposition product by hydrogen (99.9%). Figure 3.1 shows the plasma reactor system consisting of a plasma generator with a downward torch, a reactor, a cooling system, a powder feeding system, a powder collector, and an off-gas exhaust system. The plasma torch consisted of a water-cooled tungsten cathode and a copper anode nozzle operating at atmospheric pressure. The reactor consisted of a vertical water-cooled stainless steel tube of 15 cm inner diameter and 60 cm height and an inner graphite cylinder of 7.6 cm inner diameter and 60 cm length. Graphite felt was placed between the graphite tube and the inner wall of the water-cooled stainless-steel tube for insulation. The quenching chamber connected to the bottom of the reactor was a water-cooled two-layer stainless-steel box to cool the outgoing gas to a temperature below 150°C.

The powder feeding system [Ryu et al., 2008a; 2008d; 2009b] consisted of an entrained-flow powder feeder, a vibrator, a carrier gas line, a sample container, and a water-cooled delivery line. The precursor powder was placed in a glass tube to be mounted on the powder feeder. The carrier gas flowing through the glass tube fluidized the top layers of the powder in the tube. At the same time, the glass tube was gradually pushed upward at a desired rate by a syringe pump to carry the powder through the delivery line. The assembly was constantly vibrated to maintain a uniform gas-particle flow. A stainless steel tube of 3.2 mm inner diameter was used as the delivery line connected from the powder feeder to the reactor. The precursor powder was fed to the outside boundary of the visible plasma flame (7 mm diameter) from a distance of 15 mm near the exit of the plasma torch. The input amount was determined by measuring the weight of the powder in the glass tube before and after each run. Calibration of the powder feeder indicated that a constant feed rate was maintained from the beginning until all the powder was discharged.

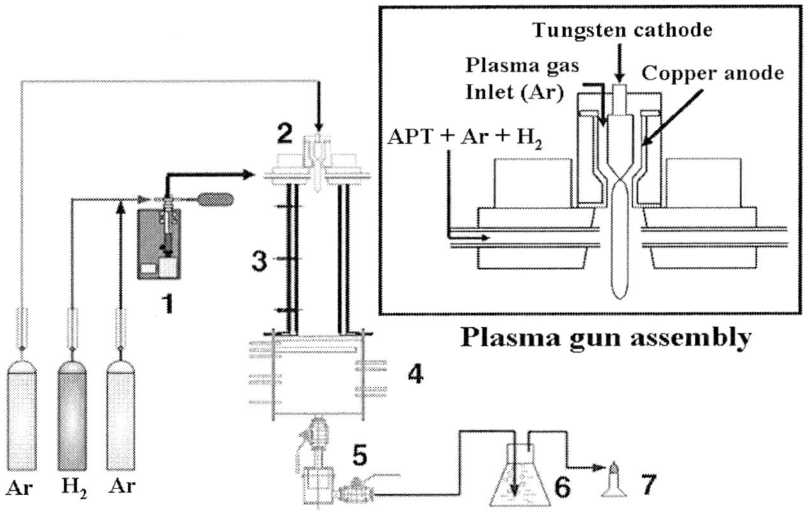

Figure 3.1. Schematic diagram of the plasma reactor system: (1) entrained-flow powder feeder for APT, (2) plasma gun, (3) cylindrical reactor, (4) cooling chamber, (5) powder collector, (6) scrubber, and (7) Bunsen burner. [Adapted from Ryu et al. (2009a)].

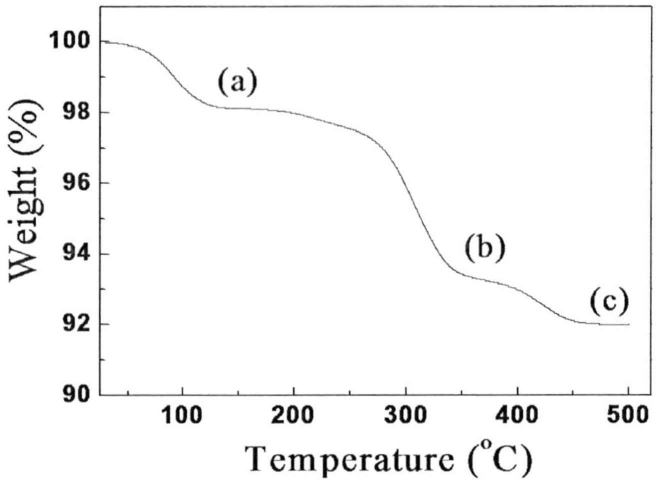

Figure 3.2. TG curve of the APT decomposition in argon (Heating rate 5°C/min). [Adapted from Ryu et al. (2009a)].

Argon (99.9%) was used as the plasma gas and a mixture of Ar and H_2 flowed through the powder feeding system to carry ammonium paratungstate into the plasma flame. The product powder was collected using a Teflon-coated polyester filter with a pore size of 1 μm.

Decomposition of APT was first examined by thermogravimetry (TG, Shimadzu TGA 50). Figure 3.2 shows the TG curve for APT in Ar, and Figure 3.3 shows the XRD patterns of the products at different temperatures during the TG runs. During the heating of APT, the water of hydration was first lost, as shown by step (a) in Figure 3.2 to anhydrous APT, corresponding to Figure 3.3(b). NH_3 and H_2O were then lost, as shown by step (b) in Figure 3.2. The solid first changed to an unknown amorphous compound, as shown in Figure 3.3(c), and then to incompletely crystallized WO_3, as shown in Figure 3.3(d), followed by fully crystallized WO_3 phase indicated by Figure 3.3(e).

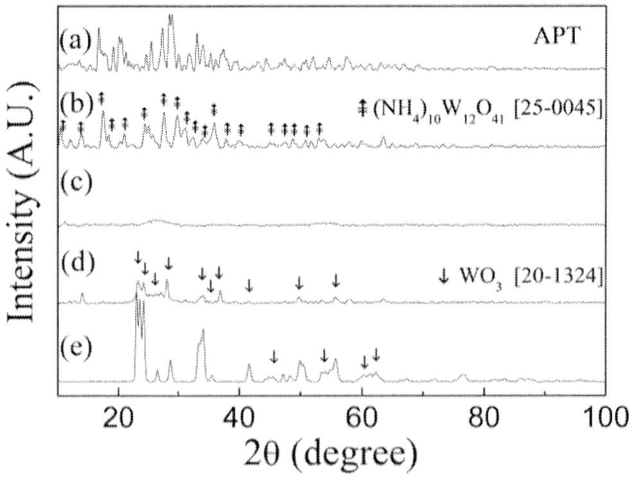

Figure 3.3. X-ray diffraction patterns of products from APT at different temperature during the TG run: (a) ammonium paratungstate as received, (b) 150°C, (c) 250°C, (d) 350°C, and (e) 500°C for 1 hour of holding time under Ar atmosphere. [The numbers after the phases are the Joint Committee for Powder Diffraction Standards (JCPDS) file numbers]. [Adapted from Ryu et al. (2009a)].

Figure 3.4. SEM micrograph of as-received ammonium paratungstate. [Adapted from Ryu et al. (2009a)].

APT was used as received and its particle size was about 40 μm, as shown in Figure 3.4. When APT powder was delivered into the plasma flame with Ar as the carrier gas, WO_3 was produced as expected from the result of heating APT in Ar atmosphere. Tungsten powder was then produced by the addition of H_2 gas as the reducing agent in the feed stream together with APT powder.

Equations given below represent the process of thermal decomposition of APT, and subsequent hydrogen reduction in the thermal plasma process [Ryu et al. 2009b]:

– Thermal decomposition

$$(NH_4)_{10}W_{12}O_{41}\cdot xH_2O \rightarrow 12WO_3 + 10NH_3(g) + (5+x)H_2O(g) \quad (3\text{-}3)$$

– Hydrogen reduction

$$WO_3 + 3H_2(g) \rightarrow W + 3H_2O(g) \quad (3\text{-}4)$$

Product Characterization

EDS analysis of the product powder exhibited only W peaks, as shown in Figure 3.5, and the XRD pattern, although not shown here, also indicated only the W phase.

The grain size of W, determined by applying the Scherrer equation to the XRD data [Cullity, 1978], was 25 ± 1 nm when the H_2 flow rate was 1.4 times the stoichiometric level and decreased to 20 ± 1 nm as the ratio was increased to 4.3. The Scherrer equation relates the XRD peak width and crystallite size, as follows:

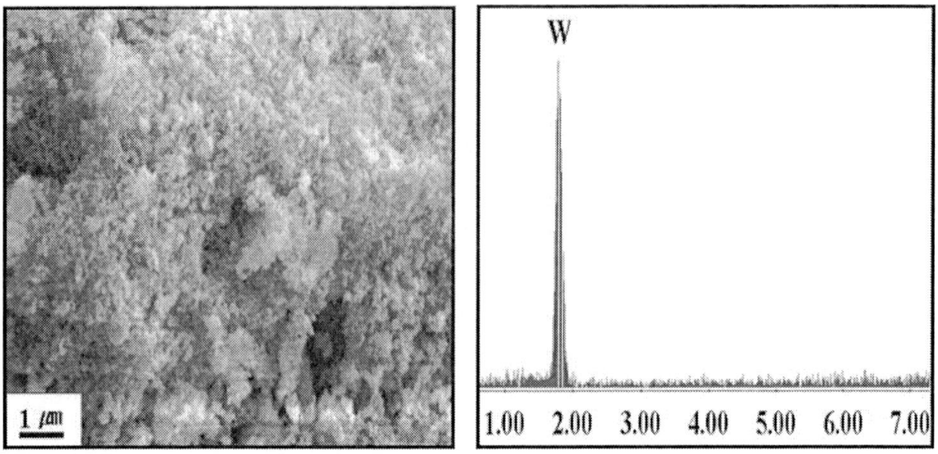

(a) (b)

Figure 3.5. SEM micrograph and EDS result of the W nanopowder obtained at the plasma torch power of 13 kW and H_2 feeding rate of 2.9 times the stoichiometric value: (a) SEM micrograph and (b) EDS result. [Adapted from Ryu et al. (2009a)].

$$Crystallite\ Size = \frac{K\lambda}{(B - B_{ex})\cos\theta} \tag{3-5}$$

where K = Shape factor (0.89 – 1.0 taken as 0.9 in this work)
λ = X-ray wavelength (1.542 Å for Cu-Kα radiation)
B = Full width at half peak height (radians)
θ = Peak position

B_{ex} = Extra broadening due to microstrain and instrumental effects (radians), obtained by the use of a standard sample.

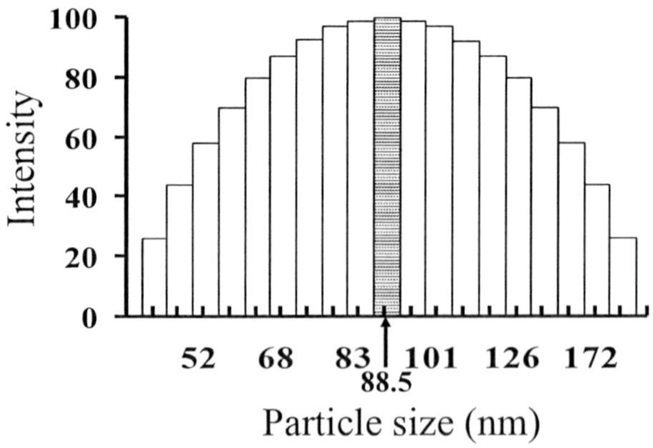

Figure 3.6. Particle size distribution in log-normal scale from ZetaPALS measurement of W nanopowder obtained at the plasma torch power of 13 kW and H_2 feeding rate of 2.9 times the stoichiometric value. [Adapted from Ryu et al. (2009a)].

It is noted that because the powders in this study were produced in vapor phase, no lattice strain effects were considered in the grain size analysis using XRD.

The particle size of W, as opposed to the grain size, was examined using ZetaPALS unit (Brookhaven Instruments) and TEM (Tecnai F30 TEM). The ZetaPALS unit with a spatial resolution of 1 nm measures particle sizes based on the degree of spectral broadening induced in an incident laser beam by scattering from particles in a Brownian motion. ZetaPALS gave an average particle size of 88.5 nm, as shown in Figure 3.6 but TEM examination showed that the actual particle size of W was less than 45 nm, as shown in Figure 3.7. The difference in the particle sizes obtained by the two methods can be explained by the agglomeration of the particles as well as their motion in liquid media that contributed to errors in particle size measurement by ZetaPALS. The particle shape was mostly spheroidal and the grain size of W did not change with plasma torch power or plasma gas flow rate within the ranges tested.

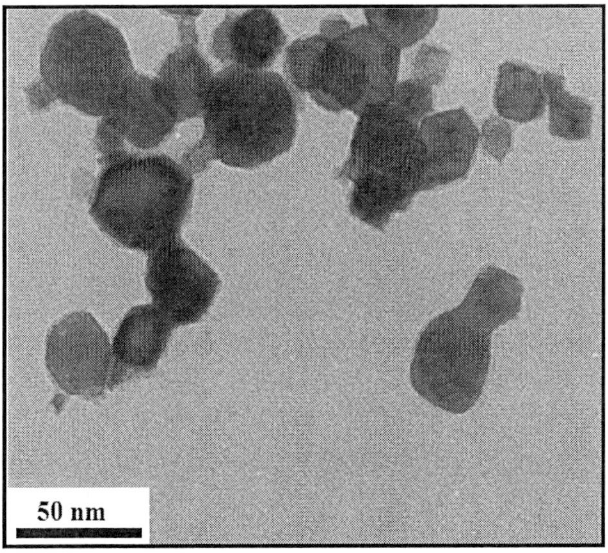

Figure 3.7. TEM photograph of W nanopowder synthesized at the plasma torch power of 13 kW and H2 feeding rate of 2.9 times the stoichiometric value. [Adapted from Ryu et al. (2009a)].

Post-Treatment

The powder obtained from the plasma process was subjected to hydrogen heat treatment to convert the minor amounts of WO_2 or WO_3 completely to W [Ryu et al., 2009a], which is required before the subsequent consolidation process. The surface of W particles can be oxidized while being exposed to air or handled, even though completely reduced W powder is obtained from the plasma process. The effect of treatment temperature on the product composition and grain growth of W was investigated. WO_2 and WO_3 phases disappeared when the powder was treated above 700°C for 1 hour, as shown in Figure 3.8. The average grain size of W, 20 ± 1 nm before the treatment, increased to 23 ± 1 nm when the powder was treated at 900°C. Figure 3.9 shows the TEM micrographs of the product treated at 700°C and 900°C for 1 hour holding time. The largest particle size of W obtained at 700°C was about 90 nm but most particles were smaller than 50 nm. When the treatment temperature increased to 900°C, the particles size of W increased to about 100 nm. In addition, the particles agglomerated more. Therefore, it was concluded that the post-treatment temperature of produced W powder should be lower than 700°C to prevent the rapid grain growth of W particles.

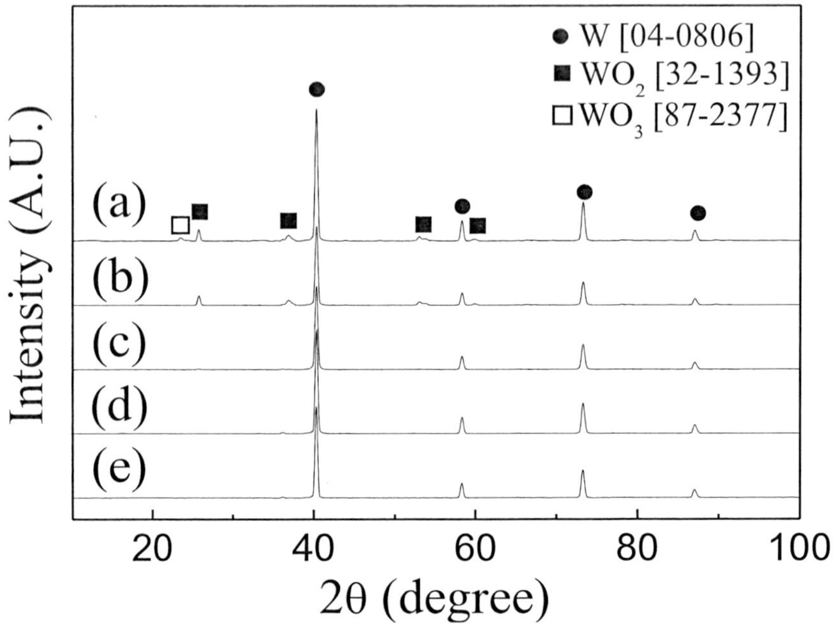

Figure 3.8. X-ray diffraction patterns of the products obtained with different hydrogen treatment temperatures using powders produced by thermal plasma process: (a) before treatment, (b) 600°C, (c) 700°C, (d) 800°C, and (e) 900°C for 1 hour of holding time. [The numbers after the phases are the JCPDS file numbers.] [Adapted from Ryu et al. (2009a)].

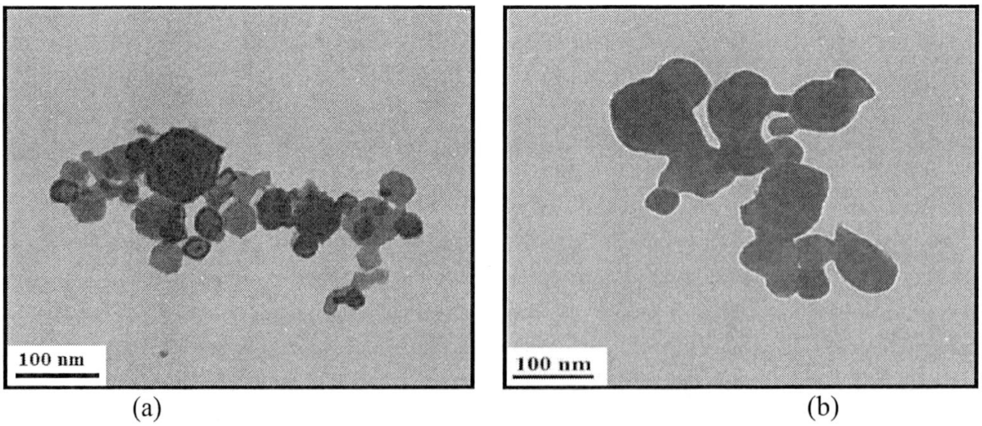

Figure 3.9. TEM micrographs of W nanopowder obtained after H_2 heat treatment at (a) 700°C and (b) 900°C for 1 hour of holding time. [Adapted from Ryu et al. (2009a)].

In summary, nanosized tungsten powder was obtained from APT by the chemical vapor synthesis (CVS) process in a thermal plasma reactor. The grain size of tungsten was less than 25 nm and was not affected by the plasma torch power and the plasma gas flow rate. But with an increase in the hydrogen flow rate, the grain size of tungsten became smaller. The produced tungsten powder was treated by hydrogen during which the minor amounts of WO_2 or WO_3 were reduced to tungsten. Finally, nanosized W powder consisting of round particles of less than 50 nm was obtained. The thermal plasma process performed in this work showed that it is an efficient technique compared with other conventional multi-step processes owing to the high temperature generated by the plasma flame to rapidly volatilize APT powder and the rapid quenching of the reduced product to directly yield nanosized W powder.

3.3. Aluminum Powder by Magnesium Reduction

Aluminum nanopowder has the highest potential as a component in a large number of hydrogen-storage materials, as described below, but is also used in many other applications such as new fuel and propellant systems [Manikama et al., 2011]. It gives some of the new fuels a higher burning rate and greater maneuverability and thrust [Kearns, 2004; Lessard et al., 2001; Mench et al., 1998; Ivanov and Tepper, 1997]. Aluminum nanopowder reacts vigorously with water which can be exploited to propel naval vehicles and ordnance, and has the potential for military pyrotechnics and detonators taking advantage of the termite reaction [Lowe, 2002; Brousseau and Anderson, 2002; Brousseau and Cliff, 2001]. Aluminum nanoparticles may be useful in the long run due to the abundance of the Al element and its relatively cheap costs compared to other elements.

Aluminum is a component in many promising hydrogen-storage materials such as aluminum hydride and complex aluminum hydrides. Hydrogen is one of the most promising alternative energy carriers [Greenwood and Earnshaw, 1998; Winter and Nitsch, 1988]. Currently, the simplest and most common method of hydrogen storage uses pressurized tanks. Other methods such as liquid hydrogen in a cryogenic tank, physisorption, metal hydrides and complex hydrides as well as the use of hydrocarbons have been studied [Sandi, 2004; Züttel, 2004; Züttel, 2003; Schlapbach and Züttel, 2001; Greenwood and Earnshaw, 1998; Winter and Nitsch, 1988]. For automotive application, numerous aluminum-containing compounds such as $NaAlH_4$ [Bogdanovic and Schwickardi, 1997; Íñiguez and Yildirim, 2005; Jensen et al., 1999], AlH_3 [Sandrock et al., 2006], $LiAlH_4$ [Balema et al., 2000] and $Mg(AlH_4)_2$ [Fichtner and Fuhr, 2002] have been identified to have high potential as hydrogen storage materials. Aluminum hydride or alane, AlH_3, is potentially a highly attractive storage material due to its high mass hydrogen capacity of 10 wt%. However, experiments on AlH_3 exhibited slow H_2 evolution rates below 150°C. It is very difficult to regenerate the spent aluminum powder back to AlH_3 [Graetz and Reilly, 2005]. One of the most promising candidates to date is the aluminum based complex hydride $NaAlH_4$. Bogdanovic and Schwickardi [1997] showed that $NaAlH_4$ can be made reversible at relatively low temperatures and pressures by adding titanium. Recently, Lu et al. [2006] discovered a combined hydrogen storage system of $LiAlH_4$-$LiNH_2$ which has a 7.3 wt% hydrogen capacity. They showed that this system is nearly 100 % reversible according to the following reaction:

$$Li_3AlH_6 + 3LiNH_2 \leftrightarrow Al + 3Li_2NH + 4.5H_2 \quad (3\text{-}6)$$

Magnesium alanate, $Mg(AlH_4)_2$, is another attractive complex hydride, which has a theoretical hydrogen storage capacity of 9.3 wt.%. Researchers [Fichtner et al., 2003; Fossdal et al., 2005; Gremaud et al., 2005; Kim et al., 2006; Komiya et al., 2007] have shown that it decomposes in two major steps according to the following reactions:

$$Mg(AlH_4)_2 \rightarrow MgH_2 + 2Al + 3H_2 \tag{3-7}$$

$$MgH_2 \rightarrow Mg + H_2 \tag{3-8}$$

$$3/2Al + Mg \rightarrow 1/2Al_3Mg_2 \tag{3-9}$$

Although metal hydrides and complex hydrides offer a safe way to store hydrogen, the kinetics of hydrogen release is too slow at a near-ambient temperature and the reactions are not readily reversible under reasonable conditions [Sandi, 2004; Schlapbach and Züttel, 2001]. To overcome these barriers, nanoparticles and doping elements have been considered. Recent studies have shown that the dehydrogenation temperature can be lowered and the reversibility of hydrides such as sodium alanate can be increased by doping the compounds with Ti [Bogdanovic and Schwickardi, 1997; Íñiguez and Yildirim, 2005; Jensen et al., 1999]. In addition, nanoscaled materials offer many advantages. The small size of these materials strongly enhances the kinetics of hydrogenation and dehydrogenation by increasing the reaction and diffusion rates. In other words, the reactivity per unit mass of nanosized materials is significantly higher due to the large surface area and the diffusion path from surface to bulk is shorter than in larger particles.

Despite the significant research on the applications of aluminum nanopowder in various areas, as discussed above, and a large number of synthesis methods tested, its production technology has not been well established and most of the proposed processes involve expensive reagents and complex steps [Manikama et al., 2011]. Light metals and their alloys are difficult to mill due to their ductility and reactivity. Therefore, chemical vapor synthesis (CVS) is one of the most suitable methods for the synthesis of Al and $TiAl_3$-containing Al nanopowders.

Recognizing this, Choi et al. [2010] carried out an investigation on the synthesis of Al and $TiAl_3$-containing Al nanopowders by the CVS process using $AlCl_3$ with Mg as the reducing agent. $TiAl_3$ is used in many Al-containing hydrogen storage materials as a catalyst for dehydrogenation reaction. The thermodynamic feasibility of this reaction has been discussed in Section 2.1.

The shape of the particles was spherical with an average size in the range of 10 ~ 50 nm measured by SEM, TEM, BET and ZetaPALS compared with the typically larger than 100 nm for commercially available fine Al powders.

In addition, the effects of the operating conditions such as Ar flow rate, precursor feed rate and reaction temperature on the properties of the product powder were investigated.

Experimental Work

Choi et al. [2010] started with preliminary experiments to examine the possibility of synthesizing Al powder by CVS. Reactor systems heated by electrical heating elements were used, each of which consisted of a tubular reactor, a volatilizer, a powder collector and an off-gas scrubber. In the preliminary tests in a vertical alumina reactor, fine powders of the reactants $AlCl_3$ and Mg were fed into small alumina tubes placed in the top part of the reactor, which served as the volatilizers. The product powder was collected in a vessel containing ethanol and the off-gas was neutralized by a NaOH solution. The collected powder was rinsed with 200-proof ethanol to remove unreacted $AlCl_3$ and $MgCl_2$ produced from the reaction, and then filtered through a filter with a pore size of 100 nm. Additional details can be found elsewhere [Choi, 2009].

The size, morphology and chemical composition of the powders were determined with SEM (TOPCON SM-300), EDS and TEM (FEI Tecnai 30). XRD (Siemens D 5000) was used to identify the phase and calculate the average grain size of the product powder. The particle size distribution was measured with a ZetaPALS unit, described in Section 3.2. The powder was suspended and dispersed with pure ethanol in a vial, which was sonicated in an ultrasonic bath for 30 minutes to minimize agglomeration of the particles.

The reactor system was subsequently modified to obtain finer nano grains with a more uniform size distribution and increase the yield of the product by minimizing the buoyancy effect of the gas phase by using a horizontal reactor [Choi et al., 2010]. Figure 3.10(a) shows the modified CVS system. The diameter of the reactor was 5 cm and a graphite funnel was installed inside the reactor to enhance the mixing of the vaporized reactants. The system was composed of external entrained-flow powder feeders, a reactor with volatilizer tubes and a powder collecting system. The volatilizers were alumina tubes that extended into the mid-part of the reactor. The powder feeders were the same as the device described in Section 3.2. $AlCl_3$ and Mg powders were placed in 1.3 and 1 cm diameter glass tubes, respectively, in a glove box and weighed. The tubes were then mounted on the powder feeder. Argon gas, shown by streams *a* and *b* in Figure 3.10(b), was flowed through the powder feeder to act as a carrier for the solid reactant as well as to maintain an inert environment. At the same time, the glass tube was gradually pushed upward at a desired rate by a syringe pump to feed the powder into the volatilizer. The flow rate of the carrier Ar was 1 $L \cdot min^{-1}$ in each stream. Additional carrier Ar at a flow rate of 0.3 $L \cdot min^{-1}$ was supplied to prevent the solid particles from sticking in the feed tubes. These flow rates were kept the same in different runs because they were the optimum rates for the feeding of the solid reactants into the reactor. Dilution Ar gas, shown as stream *e* in Figure 3.10(b), was directly supplied into the reactor at various flow rates to control the gaseous reactant concentration. The total Ar flow rate was defined as the sum of all the flow rates (carrier, additional carrier and dilution Ar flow rates) to the reactor.

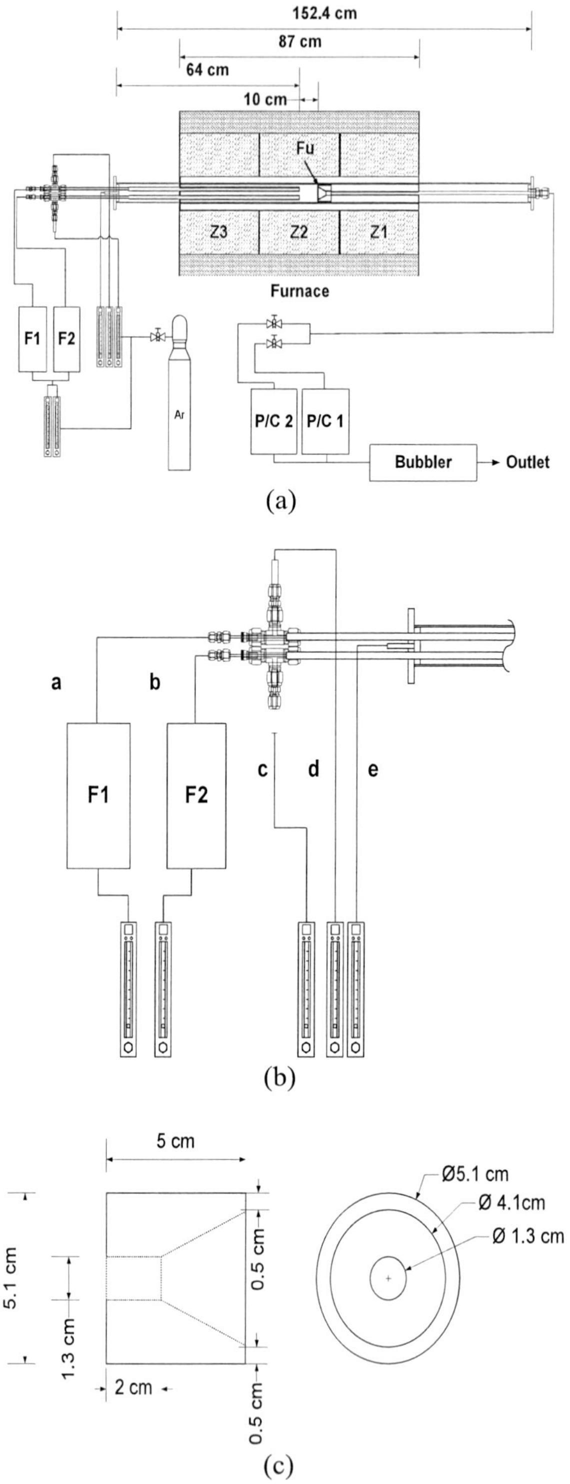

Figure 3.10. (a) Modified horizontal experimental apparatus (F1, 2: Entrained flow powder feeders, Fu: Graphite funnel, Z1, 2, 3: Hot zones, P/C1, 2: Powder collectors), (b) Detailed view of the Ar inlet area (a, b: carrier Ar, c, d: additional carrier Ar, e: dilution Ar), and (c) Detailed view of the graphite funnel. [Adapted from Choi et al. (2010)].

The graphite funnel as shown in Figure 3.10(c) was placed 10 cm away from the alumina feeding tubes to increase gas phase mixing. An alumina tube of 0.6 cm inner diameter was connected to the graphite funnel to constrict the flow and thus improve mixing of the gaseous reactants. When Ti-doped Mg powder was to be synthesized, a small amount of $TiCl_3$ particles were added to $AlCl_3$ and the two components were uniformly mixed using a mortar and pestle to be fed into the reactor. The produced powder was collected in the main filter and the off-gas from the filter was neutralized by a NaOH solution and vented.

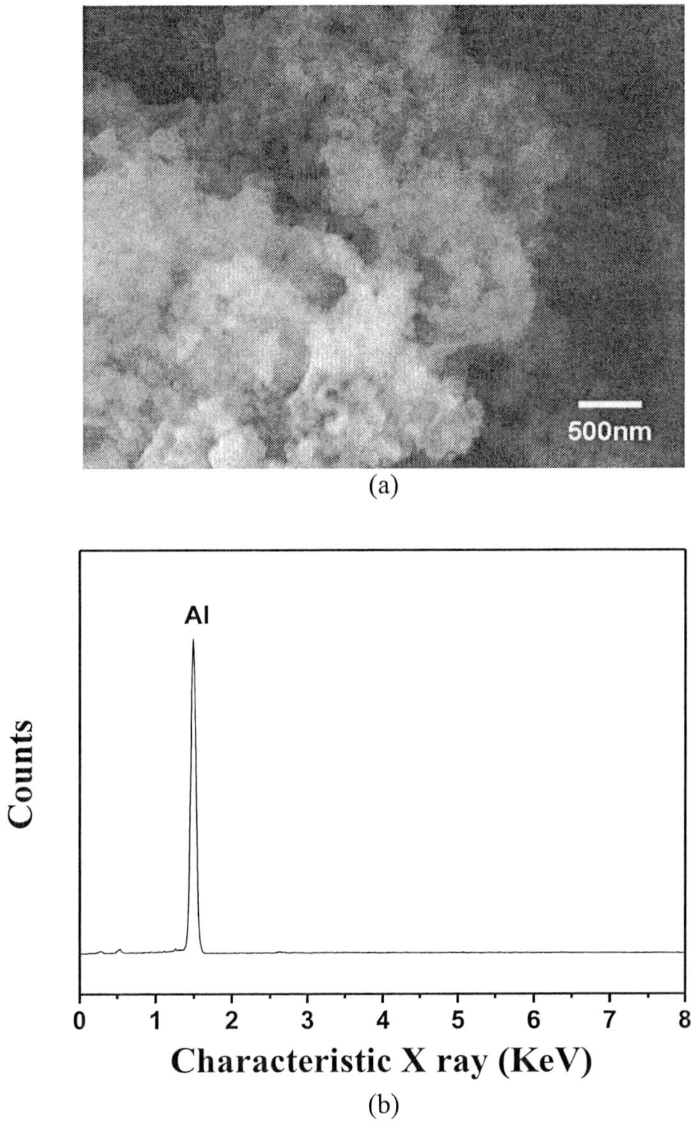

Figure 3.11. (a) SEM micrograph and (b) EDS result of Al powder produced in the improved system. [Adapted from Choi et al. (2010)].

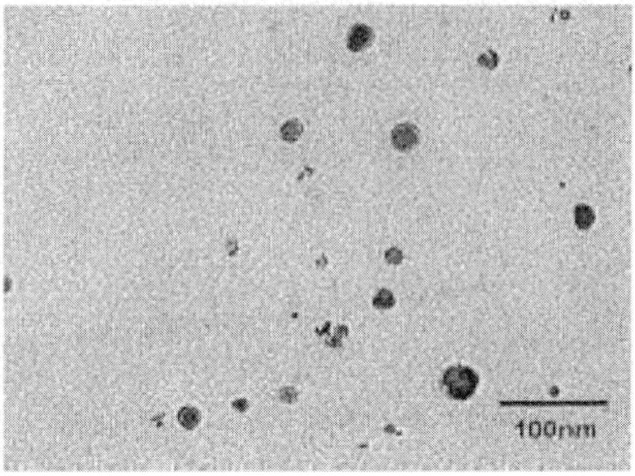

Figure 3.12. TEM micrograph of Al powder produced in the improved system. [Adapted from Choi et al. (2010)].

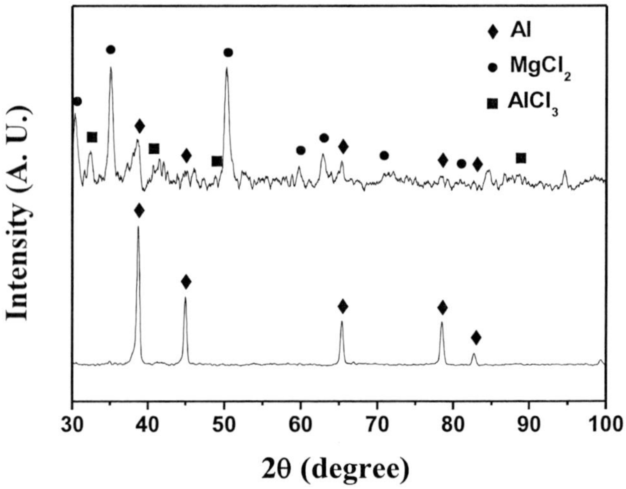

Figure 3.13. XRD patterns of product Al powder (before and after rinsing with ethanol). [Adapted from Choi et al. (2010)].

Results

Preliminary experiments in a vertical alumina tube reactor yielded nanograined Al powder with negligible amounts of impurities. The shape of the powder was spherical. The particle size was between 100 nm and 1 μm. The crystallite size calculated by the Scherrer's equation was 40 nm.

Although pure nanograined Al powder was successfully synthesized in the preliminary experiments, it was desired to obtain finer nano-grains with a more uniform size distribution.

Therefore, a new horizontal CVS system, shown in Figure 3.10, was fabricated. In this system, the feed rates of the solid reactants were accurately controlled by the use of entrained-flow powder feeders described above. Further, the horizontal system increased the product yield by reducing the loss on the wall caused by a buoyancy effect observed in the vertical apparatus. The experimental conditions were: reaction temperature of 1000°C, evaporator temperatures of 1000°C, $AlCl_3$ feeding rate of 0.1 g·min^{-1}, Mg feeding rate of 0.03 g·min^{-1} (Mg/$AlCl_3$ molar ratio = 1.4 ~ 1.6) and dilution Ar flow rate of 9.85 L·min^{-1}. Figures 3-11(a) and (b) show, respectively, the SEM micrograph and chemical composition of the synthesized powder. Only Al was detected with no other impurities in the EDS analysis. It is seen that the product powder was significantly agglomerated because of the fineness of the particles. Thus, TEM analysis was performed by diluting and sonicating the product powder in pure ethanol using a Cu grid. The TEM micrograph of the dispersed powder in Figure 3.12 indicates a spherical particle shape with sizes in the range of 10 ~ 50 nm. The particles were much smaller and more uniform than those from the previous experiments.

Figure 3.13 shows XRD patterns of the product powder before and after rinsing with the rinsed sample on the lower pattern displaying only Al peaks. Based on the XRD and EDS analysis, it was concluded that sufficiently pure Al powder was synthesized using the CVS system.

The particle size of the produced powder was also analyzed by the use of the ZetaPALS unit described in Section 3.2. Each measurement was repeated three times. Figure 3.14 presents the particle size distribution obtained, indicating an average size of 17 nm with a narrow distribution. The surface area of the particles was further measured using a BET surface area analyzer (ASAP 2100). The surface area was 103 m^2·g^{-1} corresponding to an average particle size of 22 nm. Thus, there was a good agreement among ZetaPALS, TEM and BET results in terms of the particle size.

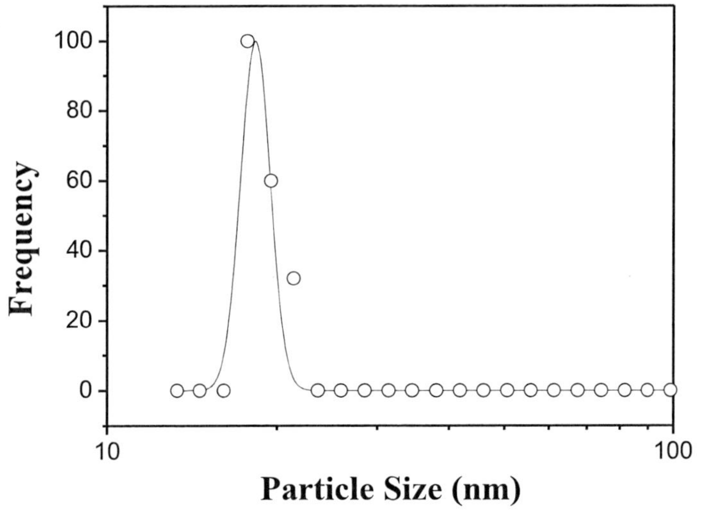

Figure 3.14. ZetaPALS particle size analysis result of product Al powder. [Adapted from Choi et al. (2010)].

The morphology of fine particles formed from vapors depends on the characteristic time of their collision and fusion during their growth. If the characteristic time for collision is much shorter than the characteristic time for fusion or sintering (e.g. $\tau_f/\tau_c > 10$), particle agglomeration will occur. If the characteristic time for collision is much longer than the characteristic time for fusion (e.g. $\tau_f/\tau_c < 0.1$) the particle will be spherical [Kodas and Hampden-Smith, 1999]. The characteristic time for sintering Al nanoparticles is of the order of pico-seconds at 800 K [Raut et al., 1998]. Because of the fast sintering time, spherical Al nanoparticles were obtained in the CVS system.

Effect of reactant concentration: The effect of reactant concentration on the Al particle synthesis was evaluated by changing the total Ar flow rate. Two alumina reactors with inner diameters of 7.9 and 5 cm were used to compare the particle size produced at different reactant concentration. The feed rates of the input precursor and reductant were kept the same ($AlCl_3$ feeding rate of 0.1 g·min^{-1} and Mg feeding rate of 0.03 g·min^{-1}). The dilution Ar flow rate was varied from 2.6 to 9.85 L·min^{-1} and the corresponding total Ar flow rate was from 5.2 to 12.5 L·min^{-1}.

Figure 3.15 shows the variation in the average particle size of product particles with the $AlCl_3$ concentration at 1000°C while the input ratio $AlCl_3$/Mg was kept the same. The particle size of the product increased with increasing $AlCl_3$ concentration. The sizes of particles formed in the larger reactor was in the range of 160 ~ 320 nm while they were 25 ~ 130 nm from the smaller reactor. The linear velocities of the gas mixture in the larger and smaller reactors were 14.5 and 36.1 cm·s^{-1}, respectively. The increase of total Ar flow rate reduced the length of time that the particles spent in the reactor and consequently led to the decrease in particle size. The lower concentration also affected the nucleation and growth of particles due to the lower degree of supersaturation.

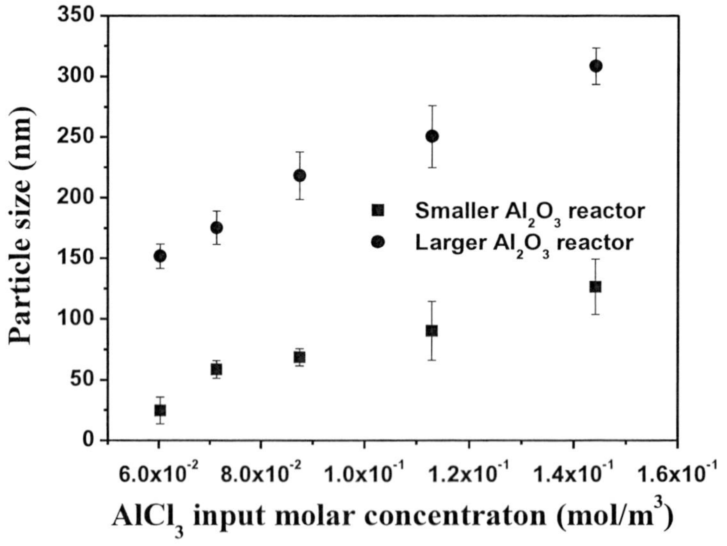

Figure 3.15. The average particle size comparison of product Al powder (at 1000°C, with various total Ar flow rates and different reactor diameters: 5 and 7.9cm). [Adapted from Choi et al. (2010)].

Synthesis of Ti-doped Al nanopowder: Titanium doping increases the dehydrogenation kinetics of most Al-containing hydrogen-storage materials, as mentioned earlier. Typically, Ti doping is done by milling, but it is difficult to mill Al powder due to its high ductility. In the CVS process, doping can readily be accomplished during the powder synthesis by directly mixing $TiCl_3$ with the precursors. Uniform distribution of the dopant on the particle surface is important to maximize its effect. Two types of doping methods were used in the work of Choi et al. [2010].

In the first method, 4 mol% $TiCl_3$ powder was premixed with $AlCl_3$ powder. This percentage was based on just the weights of $TiCl_3$ and $AlCl_3$, not including Mg. The input molar ratio of $AlCl_3$/Mg was 0.64. The dilution Ar flow of the reactor was 9.85 $L \cdot min^{-1}$ and the temperature was 1000°C. Figure 3.16 shows the SEM image and EDS mapping results of the product powder. The dopant was uniformly distributed in the products. The EDS result in Figure 3.17 shows that the CVS powder was composed of Al, Cl, Ti and a small amount of oxygen. The molar ratio Ti/Al in the product power was 0.06, based on the EDS quantitative analysis. The Ti mole fraction in the product was higher than in the feed. This is likely due to the higher ($-\Delta G°$) value of the reaction between Mg and $TiCl_3$ (-34.82 kcal at 1000°C) than that of the reaction between Mg and $AlCl_3$ (-26.11 kcal at 1000°C). Thus, $TiCl_3$ was preferentially reduced by the Mg vapor in the reactor. Further, the XRD pattern in Figure 3.17 indicates that titanium was present mostly in the form of the intermetallic compound $TiAl_3$. The $\Delta G°$ of the $TiAl_3$ formation at 1000°C is -24.8 $kcal \cdot mol^{-1}$, which makes the formation reaction favorable.

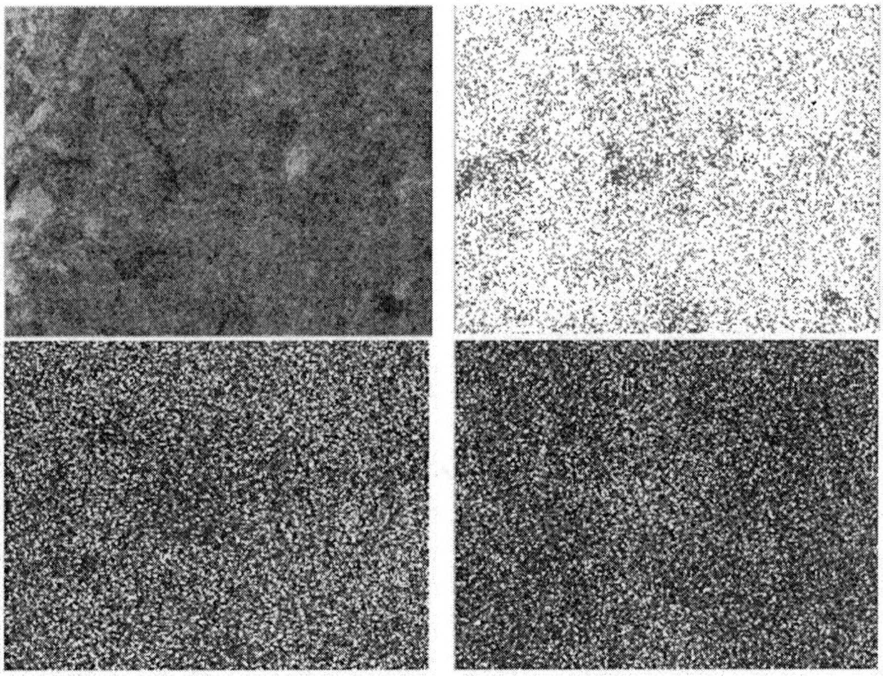

Figure 3.16. EDS mapping results of the 4 mol% $TiCl_3$ doped Al powder (white regions indicate the elements) (top left) SEM image of the powder, (top right) Al, (bottom left) Ti, and (bottom right) Cl distribution map. [Adapted from Choi et al. (2010)]

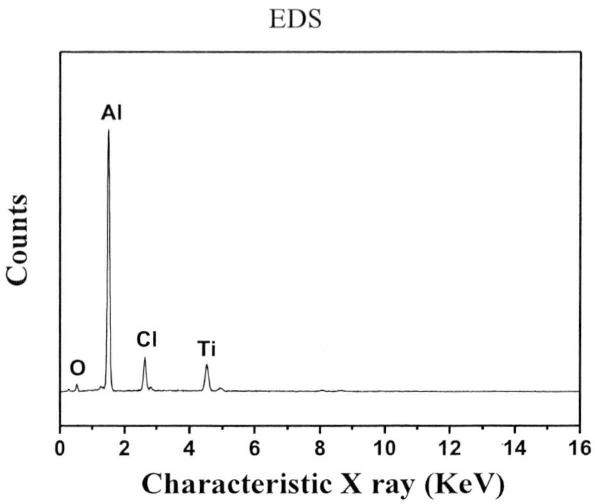

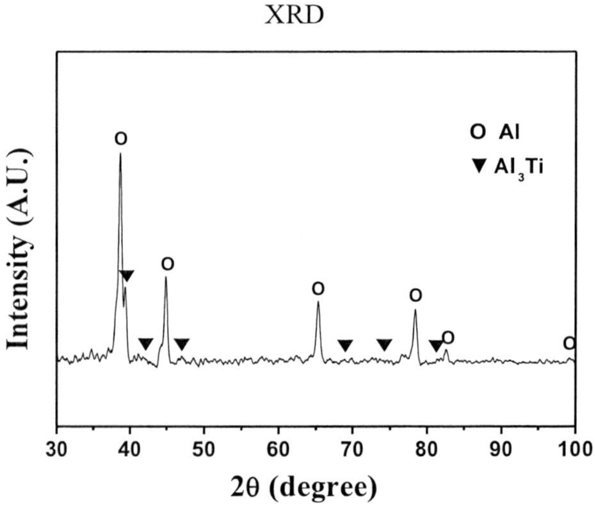

Figure 3.17. EDS analysis and XRD patterns of the Al powder from the $AlCl_3$/4 mol% $TiCl_3$ mixture. [Adapted from Choi et al. (2010)].

The intermetallic compound $TiAl_3$ has been reported to improve the dehydrogenation kinetics of $NaAlH_4$ [Graham et al., 2007; Kang et al., 2006; Bogdanovic et al., 2003; Jensen et al., 1999].

In the second method, a liquid phase doping method was applied because this method provides another efficient way for doping. Ethanol that contains 4 mol% dissolved $TiCl_3$ was mixed with the Al powder produced by the magnesium reduction of $AlCl_3$ and dried in a glove box to prevent oxidation. The EDS mapping results shown in Figure 3.18 indicate that $TiCl_3$ was uniformly distributed on the product powder. The product contained 1.0 at% Ti together with 5.4 at% Cl, based on the EDS elemental analysis of the sample shown in Figure 3.19. This method did not produce any intermetallic phases.

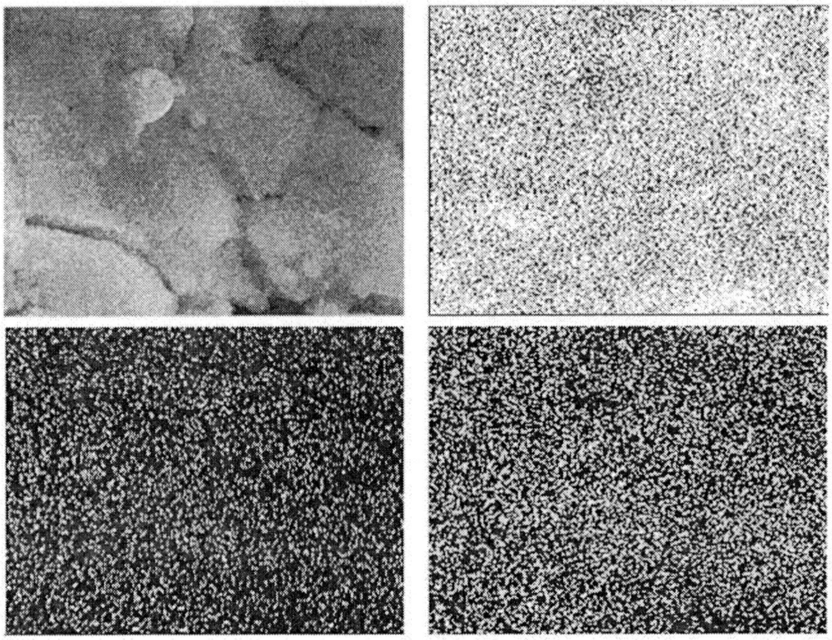

Figure 3.18. EDS mapping of the liquid phase $TiCl_3$ doped Al powder (white regions indicate the elements) (top left) SEM image of the powder, (top right) Al, (bottom left) Ti, and (bottom right) Cl distribution map. [Adapted from Choi et al. (2010)].

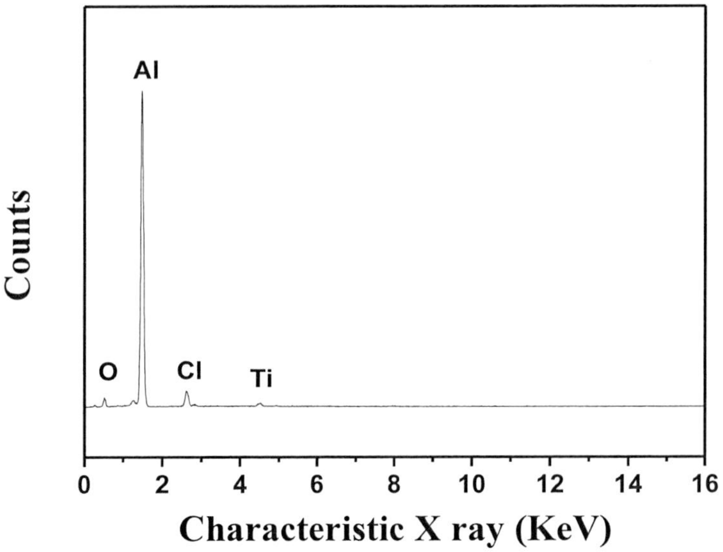

Figure 3.19. EDS analysis of the liquid phase Ti-doped CVS Al powder. [Adapted from Choi et al. (2010)].

Conclusion

The CVS technique was applied to the preparation of nanosized Al powder using a horizontal furnace and external powder feeding system, which has several advantages. Firstly, it is a combined unit that has a volatilization system and a reaction system inside the reactor.

Secondly, the powder feeding system provides an accurate control of the solid reactant feed rate. Using this CVS system, pure Al nanoparticles of ~ 20 nm size and $TiAl_3$-containing Al powders were synthesized. Uniform Ti-doping on the Al nanopowder was achieved using premixed precursors. The CVS system has also proved to be capable of producing nanosized particles with narrow size distributions.

3.4. Titanium Powder by Magnesium Reduction

Titanium is mostly produced by the Kroll process, in which titanium tetrachloride is reduced with magnesium in a steel pot [Paschen et al., 1990; Minkler and Baroch, 1981]. The process has not changed much since its first development in 1940 [Kroll, 1940], and its batch-wise nature makes it labor and energy-intensive. The process also suffers from low productivity because the reactor has to be charged, heated, and discharged, which involves a long downtime between batches and provides an ample opportunity for emission of chloride vapors.

Sohn [1998] tested a CVS process that is continuous and closed, and will virtually eliminate the possibility of emission during the reduction process. In this process, titanium powder is produced by the reduction of titanium tetrachloride by magnesium in the vapor phase by the following reaction:

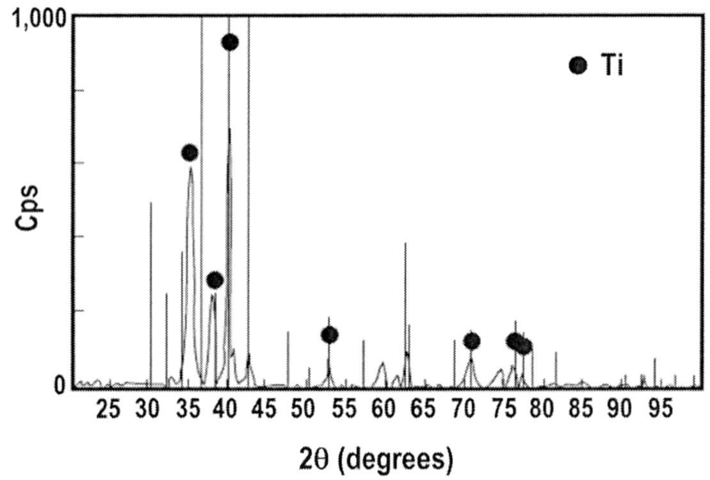

Figure 3.20. An x-ray diffraction pattern of the product powder indicating relatively pure titanium. [Adapted from Sohn (1998)].

$$TiCl_4(g) + 2Mg(g) = Ti(s) + 2MgCl_2(g); \quad (3\text{-}10)$$

$\Delta H^o = -36.5$ kcal/mol Mg at 1,400 K

The CVS process can be carried out in a continuous reactor by feeding the reactants into a high-temperature environment for rapid volatilization and reaction. The magnesium-reduction reaction is highly exothermic, and the reaction in the vapor phase is rapid. An additional benefit of the proposed concept is that fine particles down to nanosized powder can be produced directly.

For high-melting materials, such as titanium, the direct production of fine powders provides significant advantages in terms of the subsequent processing of these materials into their final forms. A major concern with production in the form of fine powder is its tendency to form oxide, which should be carefully avoided.

Titanium powder was produced based on the above concept in a furnace measuring 0.24 m in diameter and 1.4 m in length heated to 900-950°C. In these tests, liquid titanium tetrachloride and magnesium powder were fed continuously with a small amount of argon flow for periods of about one hour. The product powder was collected, washed with mild hydrochloric acid solution followed by water, then dried and analyzed with x-ray diffraction. The x-ray pattern of the obtained product in Figure 3.20 shows that the sample is essentially all titanium with small amounts of unreacted magnesium and possibly a trace of magnesium oxide.

3.5. Composite Nanopowders

Because particles are formed from gaseous reactants in the CVS process, very uniformly mixed composite powders can be obtained from a gas mixture that contains precursors of different solid phases. Several examples of practical importance that have verified this expectation are discussed below.

Magnesium-Titanium

Magnesium and magnesium-based alloys are attractive candidates for rechargeable hydrogen storage materials because of their high hydrogen capacities (theoretically up to 7.6 wt%). However, the potential for practical use of MgH_2 is limited because of the low rates of its hydrogen release/uptake reactions, especially at temperatures below 300°C [Grochala and Edwards, 2004]. MgH_2 is thermodynamically highly stable in the temperature range of practical interest. The decomposition temperature of MgH_2 at 1 bar H_2 pressure is 275°C, which is too high for vehicular applications. Thus, it is important to develop a method to lower the dehydrogenation temperature of MgH_2 for practical applications. In recent years, much research on MgH_2 has focused on reducing the particle size or doping it with catalysts [Choi et al., 2008; Kyoi et al., 2004; Oelerich et al., 2001; Zaluska et al., 1999].

Magnesium nanomaterials synthesized by gas phase reaction have been reported to possess especially distinct hydrogen storage properties [Shao et al., 2004; Gleiter, 2000]. Sohn and coworkers [Choi et al., 2008; 2009a; 2009b; Lu et al, 2009] investigated the hydrogen storage characteristics of the MgH_2-$0.1TiH_2$ systems and concluded that nanostructured MgH_2-$0.1TiH_2$ nanocomposite powder prepared by ultrahigh-energy-high-pressure reactive milling is a promising reversible hydrogen storage material with a capacity of 6 wt% hydrogen. The material demonstrated high cycle stability with little loss of capacity over 80 cycles. HRTEM and STEM analyses showed that the grain size of the milled MgH_2-$0.1TiH_2$ powder was approximately 5-10 nm with TiH_2 distributed uniformly among the

MgH_2 particles. The results indicated that both the nanosize and the addition of TiH_2 contributed to the improved kinetics of dehydrogenation and hydrogenation of MgH_2. The results also showed that the ΔH value of the dehydrogenation of MgH_2-0.1TiH_2 was lower than that of MgH_2.

Based on the above findings, Choi et al. [2009a] applied the CVS process to the preparation of nanocomposite magnesium-titanium powders. The apparatus used for this purpose is shown in Figure 3.21, which consisted of a vertical tubular reactor, an electrically heated furnace, two separate powder feeders, volatilizers, a powder collector, and an off-gas scrubber. The reactor was an alumina tube with an inner diameter of 7.9 cm and a length of 1.52 m. The external powder feeders were the same as the device described in Section 3.2, designed to deliver fine precursor powders with sizes between 200 and 400 μm into the reactor. The fed precursors were evaporated in two separate alumina volatilizers placed inside the reactor and then the vaporized metal chloride was reduced by Mg vapor. The collected powder was rinsed with pure ethanol to remove unreacted titanium chloride and the $MgCl_2$ produced from the reaction and was filtered using a vacuum filter with a pore size of 40 nm. The product powders were characterized by the use of SEM (Topcon SM-300), EDX (EDAX), TEM (FEI Tecnai 30), BET (Micromeritics ASAP 2010), and XRD (Siemens D 5000).

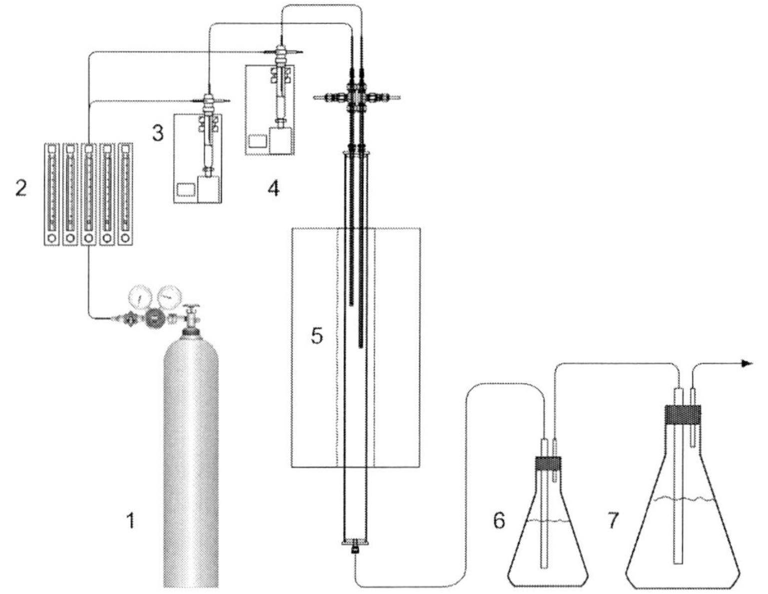

Figure 3.21. Schematic diagram of the experimental apparatus: (1) Ar gas, (2) flow meters, (3) $AlCl_3$ powder feeder, (4) Mg powder feeder, (5) furnace, (6) powder collector, and (7) scrubbing solution. [Adapted from Sohn et al. (2007)].

The overall reaction of $TiCl_3$ and Mg in the gas phase to synthesize Mg powder doped with Ti is represented by:

$$0.05TiCl_3(g) + Mg(g) = 0.05Ti(s) + yMg(g) + (1\text{-}y)MgCl_2(g) \quad (3\text{-}11)$$

The effect of temperature on the equilibrium composition in the Mg-Ti-H_2 system was examined by the use of HSC Chemistry software [Outotec, 2002], as shown in Figure 3.22. The results indicate that Mg-Ti powder can be made by rapidly quenching the product from above 1000°C. The detailed experimental conditions are described in Choi et al. [2009a].

Figure 3.23 shows the SEM micrograph and EDS mapping image of the produced powder, which was synthesized at the reaction temperature of 1450ºC. In Figure 3.23(A), the shape of the particles is seen to be spherical and considerably agglomerated. In addition, the image of EDS compositional mapping of Ti shown in Figure 3.23(B) indicates a uniform distribution of Ti among Mg particles, with a Ti content of approximately 5 at%.

The particle size was determined by BET and TEM in addition to XRD. The BET surface area data (S_{BET}) can be used to calculate the mean diameter (D_{BET}) of the equivalent spherical particles having the same specific surface area, as follows:

$$D_{BET} = \frac{6}{\rho S_{BET}} \tag{3-12}$$

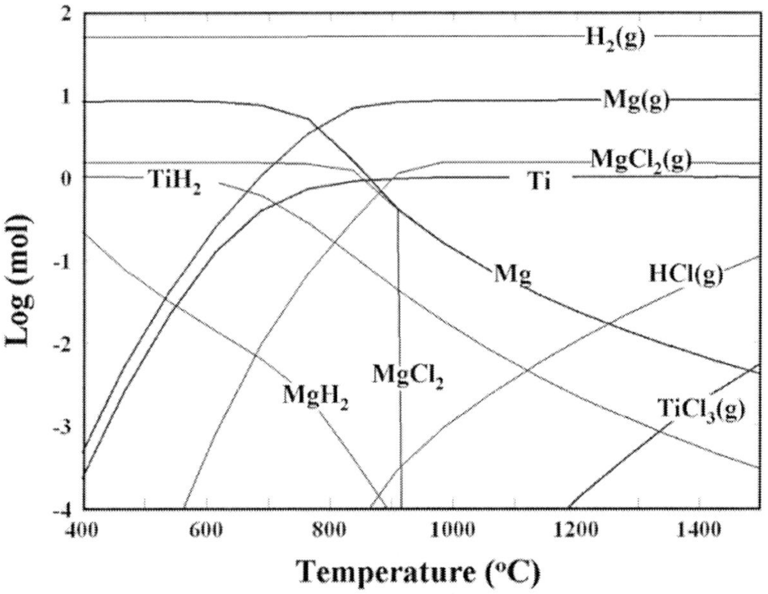

Figure 3.22. Equilibrium composition vs. temperature for a mixture of Mg-Ti-H_2 system (1 mol Mg - 0.05 mol $TiCl_3$ - 1.5 mol H_2). [Adapted from Choi et al. (2009a)].

where ρ is the theoretical density of Mg (1.74 g/cm). The specific surface area and the equivalent mean diameter of the particles were 11.2 m^2/g and 307 nm, respectively. This value is comparable to that from the TEM image shown in Figure 3.24. It indicates that the synthesized Mg-Ti powders are nearly spherical with diameters of 50 ~ 500 nm although the particles were somewhat agglomerated, as also indicated by Figure 3.23.

The X-ray diffraction pattern of the synthesized Mg-Ti powder, shown in Figure 3.25, indicates the presence of the phases Mg, MgH_2, Ti, and $TiCl_3$. MgH_2 also formed during the CVS process in this tubular reactor system, which likely happened while Mg particles were cooled. The grain size of synthesized powder was calculated from the XRD pattern using the

Scherrer equation described in Section 3.2. The crystallite size, which was around 26 nm, is smaller than that from SEM or BET analyses mainly because the latter only measures the particle size rather than the grain size, and the synthesized particles were agglomerated and/or sintered.

Magnesium-Aluminum

Aluminum has also been found to destabilize MgH_2, and forms Mg/Al alloys upon dehydrogenation [Crivello et al, 2009; Andreasen, 2008; Crivello et al., 2007; Zaluska et al., 2001; Bouaricha et al., 2000]. For example, Crivello et al. [2009] reported that Mg-Al alloy powder prepared by ball milling released 4.7 wt% hydrogen at 250ºC. Thus, Mg-Al composite nanopowders can be used as a starting material for hydrogen storage.

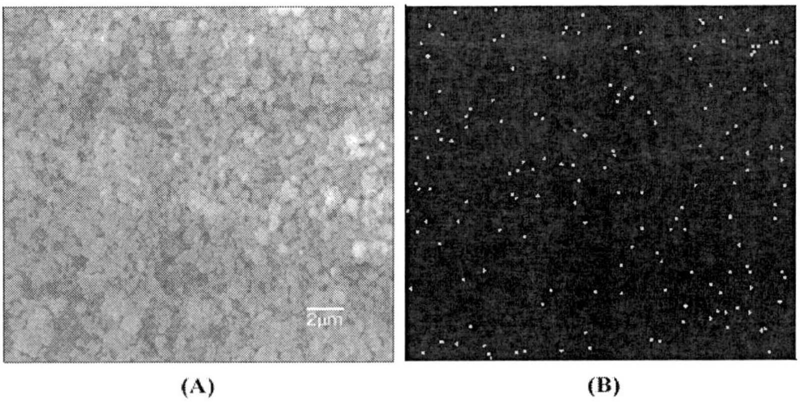

Figure 3.23. (A) SEM image of the Mg-Ti nanopowder, and (B) EDS image for the compositional mapping of Ti (white color) under the following conditions: Mg feeding rate of 0.05 g/min, $TiCl_3$ feeding rate of 0.015 g/min, hydrogen gas flow rate of 0.15 L/min (25ºC, 86.1 kPa), and total Ar flow rate at 1.5 L/min (25ºC, 86.1 kPa). [Adapted from Choi et al. (2009a)].

Figure 3.24. TEM image of the Mg-Ti nanopowder. [Adapted from Choi et al. (2009a)].

Choi and coworkers [2010] synthesized a mixture of Al and Mg by reducing $AlCl_3$ with excess amounts of Mg. The elemental compositions obtained from quantitative EDS analysis were plotted against the input ratio of $Mg/AlCl_3$. As shown in Figure 3.26, the product was essentially pure Al until the input molar ratio reached the stoichiometric value of 1.5. Beyond this ratio, the mole fraction of Al decreased proportionally with increasing Mg vapor supply. As expected from the phase diagram [ASM International, 1992], intermetallic compound Al_3Mg_2 was formed by increasing the Mg input, which was confirmed by XRD analysis (not shown).

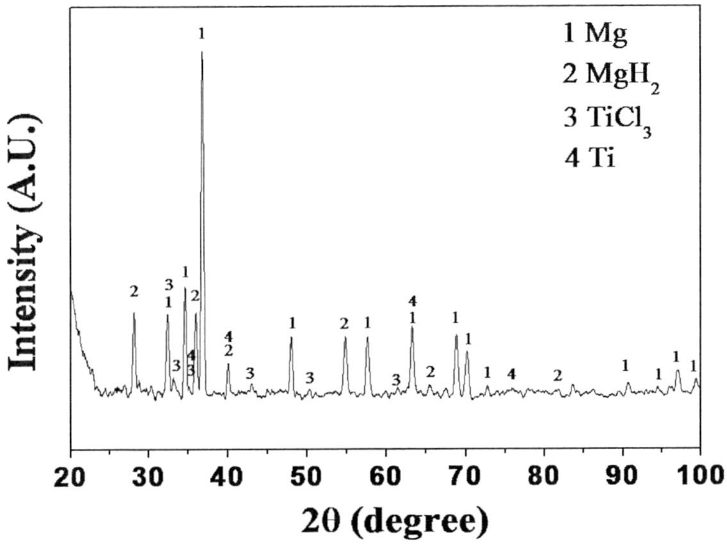

Figure 3.25. XRD patterns of the synthesized Mg-Ti nanopowder. [Adapted from Choi et al. (2009a)].

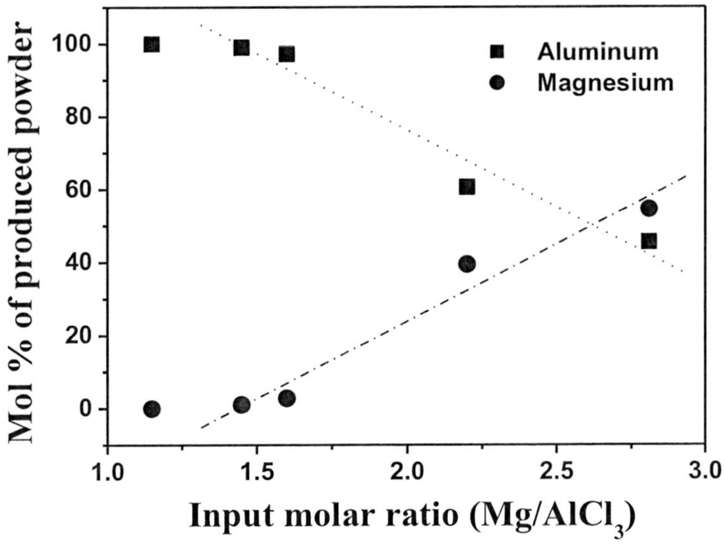

Figure 3.26. Effect of Mg vapor supply on product composition. [Adapted from Choi et al. (2010)].

Tungsten-Cobalt

Nanosized tungsten and cobalt composite powders make an excellent a raw material for the preparation of cemented tungsten carbide, as will be discussed in Section 6.2. Ryu et al. [2010a] synthesized such a composite powder by a thermal plasma process from ammonium paratungstate (APT) and cobalt oxide. The injected precursors were vaporized in a plasma flame and the subsequent hydrogen reduction of the vaporized precursors produced uniformly mixed nanosized tungsten and cobalt composite powder. The subsequent carburization of this powder by CH_4-H_2 mixture resulted in the formation of nanosized WC-Co composite powder with a particle size less than 110 nm. This new method represents an alternative route that avoids the post-treatment required for the WC_{1-x} produced when methane was used for direct carburization in the powder production step.

It is known that APT is easily decomposed to WO_3 by heating [Fait et al., 2008; Madarász et al., 2004; Medeiros et al., 2001; van Put, 1995]. The equilibrium composition diagram for the WO_3-Co_3O_4-H_2 mixture obtained by the use of HSC Chemistry software [Outotec, 2002] was shown in Section 2.3. It was indicated that it is thermodynamically feasible to synthesize W and Co from the WO_3-Co_3O_4-H_2 mixture in the plasma process. Ammonium paratungstate [$(NH_4)_{10}W_{12}O_{41}\cdot xH_2O$, 99.97%] and cobalt oxide (Co_3O_4, 99%) of 40 μm average particle sizes were used as the precursors by Ryu et al. [2010a]. The thermal plasma system and the powder feeding system used to synthesize tungsten and cobalt powder have been described in Section 3.2. Experimental conditions to synthesize W and Co composite powder were as follows: Plasma torch power was 13 kW, the flow rate of plasma gas (Ar) was 28 L/min, feeding rate of pre-mixed APT-Co_3O_4 with a Co/W molar ratio of 0.7, which was aimed to produce WC-17 wt% Co as in a typical industrial application, was 3.5 g/min and the flow rate of H_2 to carry APT-Co_3O_4 mixture was 4 L/min (25°C, 86.1 kPa total pressure at Salt Lake City).

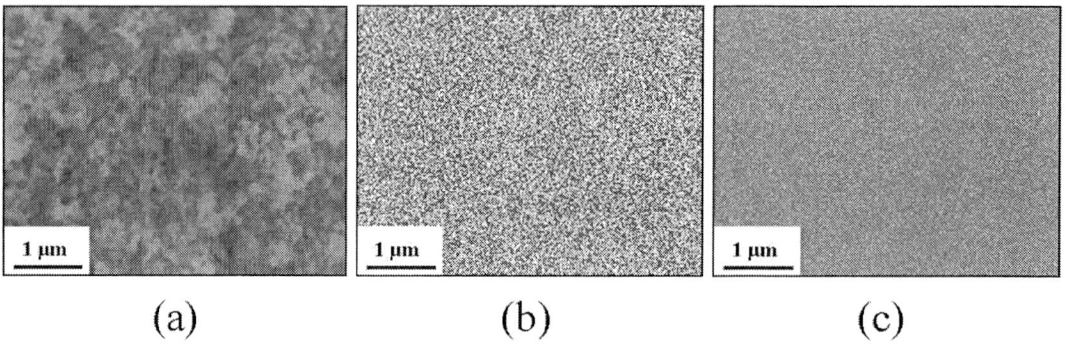

Figure 3.27. SEM image of W-Co composite powder with associated EDS maps for tungsten and cobalt: (a) W-Co composite powder, (b) EDS map of W (light specks), and (c) EDS map of Co (light specks). [Adapted from Ryu et al. (2010a)].

The synthesized W-Co powders were collected using a Teflon-coated polyester filter with a pore size of 1 μm. The morphology of the produced powder was observed using an analytical scanning electron microscope (SEM, Topcon sm-300) equipped with energy dispersive X-ray spectrometry (EDS) allowing visualization of the distribution of elements.

The morphology and particle size were also examined by the use of TEM (FEI, Tecnai F30). The grain size of the synthesized powders was calculated from the XRD pattern by applying the Scherrer equation described in Section 3.2. It is noted that a particle can be made up of polycrystalline grains or be just a single crystalline grain [Cullity 1978].

The grain sizes of the W and Co determined from the XRD data were 20 ± 1 nm and 7 ± 1 nm, respectively. The internal strain and lattice distortion effects were not taken into account in this calculation because the powder was produced in the vapor phase and thus did not experience stress. The Co/W molar ratio in the collected powders was about 0.4, which was lower than that in the input, due to the greater tendency of Co to be deposited on the reactor wall during the reaction.

In conventional reduction processes, cobalt-containing intermetallic phases of Co_3W and/or Co_7W_6 are produced from tungsten-cobalt oxide mixture. Normally, it requires several hours to complete the reduction to W-Co by this conventional method [McCandlish and Polizzotti, 1989; French and Sale, 1985]. In the thermal plasma process, no such intermetallic phases were present in the product.

The results of EDS mapping of the W-Co composite powder are shown in Figure 3.27, which indicate that W and Co particles were uniformly mixed in the product. This represents an example of the significant feature of the CVS reaction to produce very uniformly mixed powders.

Chapter 4

INTERMETALLIC AND ALLOY POWDERS

ABSTRACT

This chapter describes the application of the chemical vapor synthesis (CVS) process to the preparation of intermetallic powders by the simultaneous reduction of the constituent metal compounds. Several examples of note presented here show the significant advantages of CVS in preparing compound particles that require more than one precursor. This is due to the fact that vapor phase precursors can be mixed completely. The advantage is especially significant when preparing materials that have high melting points or otherwise difficult to handle by conventional processing techniques.

Intermetallic compounds find many applications based on their many useful properties such as low density, high melting points, high-temperature strength, and other specific properties [Liu and Kumar, 1993; Kim, 1994; Dastafani, 1989]. For example, titanium and nickel aluminides have attracted a great deal of attention from the aerospace as well as automobile industries because of their low densities, good oxidation resistances, and high-temperature strength retention [Yamaguchi and Inui, 1993]. Gamma TiAl and two-phase ($TiAl + Ti_3Al$) alloys [Kim, 1994] are used for gas turbine rotational components as well as automobile exhaust valves. $TiAl_3$ is one of the most extensively investigated aluminum-rich intermetallics. Nickel aluminides are attractive because of their low densities, relatively high melting points, and good high-temperature strength properties. For example, the yield strength of Ni_3Al increases with increasing temperature unlike most materials [Thornton et al., 1970]. A major shortcoming of the intermetallic compounds is their brittleness [Yamaguchi and Inui, 1993; Liu et al., 1983], which makes conventional casting difficult. It would therefore be advantageous if they can be produced as fine powders, especially if they are produced directly from intermediate raw materials rather than expensive and high-melting constituent metals. The powders can readily be consolidated by powder metallurgical techniques.

4.1. Titanium Aluminides by Magnesium Reduction

The current practice for producing titanium-aluminide intermetallic compounds is difficult and time consuming. It uses titanium and aluminum metals. The difficult issues associated with the current method of titanium production were discussed in Section 3.4. Aluminum is produced by the energy-intensive Hall-Héroult process. Titanium and aluminum must be melted at a high temperature due to the high melting points of the aluminides. The product aluminides are difficult to process because of their brittleness and high melting points. Thus, it would be highly desirable to develop a continuous process to produce titanium aluminides in powder form. Here, developments in the synthesis of such powders by the chemical vapor synthesis (CVS) process are summarized.

Sohn and PalDey [1998a] investigated the synthesis of titanium aluminides by the CVS method. They first carried out equilibrium calculation of the gaseous and condensed phases for the Ti-Al-Mg-Cl-Ar system. Equilibrium compositions of the solid elemental and intermetallic phases were calculated as functions of reactant partial pressures and temperature, as discussed in Section 2.1. The effects of reaction conditions on the formation of different intermetallic phases were then studied experimentally. The size and shape of the fine particles were observed by transmission electron microscopy (TEM).

Experimental Results

A schematic diagram of the quartz reactor, consisting of a gas inlet, chloride vaporizers, main reactor, and a particle collector used by Sohn and PalDey [1998a], is shown in Figure 4.1. The precursor chlorides were vaporized in separate vaporizers and were carried by argon gas to the reactor. In this work, $TiCl_3$ rather than the more readily available $TiCl_4$ was used as the input titanium chloride. This was because $TiCl_3$ has a much lower vapor pressure, which facilitated the control of its vapor pressure in a range similar to those of other reactants. A stream of argon gas was introduced at the front end of the main reactor tube to improve mixing and prevent back flow of reactant gases.

The vaporizers were made of quartz tubes of 4-cm diameter and 40-cm length. The main reactor was a quartz tube of 4-cm diameter and 150-cm length, which was tapered in the midsection to 3-cm diameter. A graphite boat containing Mg chips was pushed inside the reactor and was placed next to the mixing zone to produce Mg vapor. A layer of cotton placed on a fritted disc served as the particle collector. The remaining gas was bubbled through water. The carrier gas flow rate was 0.5 L/min at 25°C and 86.1 kPa (atmospheric pressure at Salt Lake City). The partial pressures of the reactant chlorides were calculated by measuring the vaporization rates of the chlorides at a particular temperature. After the experiment was over, the solid product was washed with dilute HCl solution followed by water to remove unreacted chlorides and was subjected to X-ray diffraction (XRD), TEM analysis, and elemental analysis by DCP (direct-current plasma spectroscopy).

The Mg vapor reduction of metal chlorides in this system formed $TiAl_3$ and TiAl intermetallic compounds along with elemental Ti and Al. The chemical reactions involved are as follows:

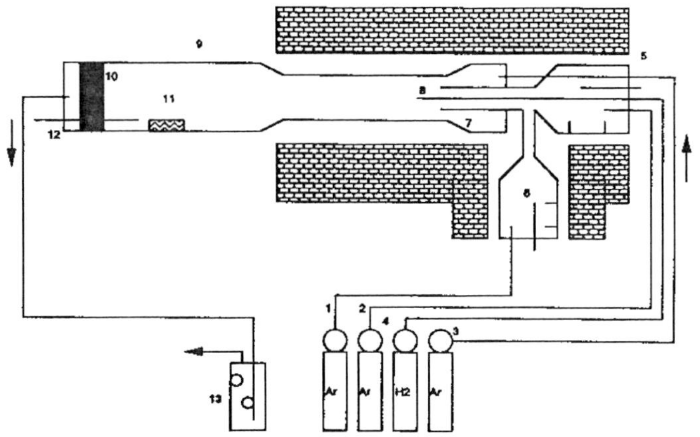

Figure 4.1. Schematic diagram of the experimental apparatus: 1, 2, 3—Ar gas, 4—H2 gas, 5, 6—chloride vaporizers, 7—mixing zone, 8—reaction zone, 9— cooling zone, 10—particle collector, 11—graphite boat, 12—quartz pusher rod, and 13—gas absorbing solution. [Adapted from Sohn and PalDey (1998a)].

$$TiCl_3(g) + AlCl_3(g) + 3Mg(g) = TiAl(s) + 3MgCl_2(g,l) \tag{4-1}$$

$$TiCl_3(g) + 3AlCl_3(g) + 6Mg(g) = TiAl_3(s) + 6MgCl_2(g,l) \tag{4-2}$$

$$TiCl_3(g) + 1.5Mg(g) = Ti(s) + 1.5MgCl_2(g,l) \tag{4-3}$$

$$AlCl_3(g) + 1.5Mg(g) = Al(l) + 1.5MgCl_2(g,l) \tag{4-4}$$

Magnesium is a strong reducing agent and thus can reduce $TiCl_3$ and $AlCl_3$ individually. Therefore, these reactions proceeded without the need of a nucleation promoter, in contrast to the hydrogen reduction of nickel chloride and aluminum chloride mixture to produce nickel aluminides, as will be further discussed in Section 4.3 [Sohn and PalDey, 1998d]. The solid product particles always contained one or two intermetallic phases and a metal. In the cases where the product consisted of two phases, e.g., $TiAl_3$ and Al, the amount of each phase was calculated as follows:

$$moles\ of\ TiAl_3 = atoms\ of\ Ti \tag{4-5}$$

$$atoms\ of\ Al\ in\ elemental\ Al = total\ number\ of\ Al\ atoms - 3\ X\ atoms\ of\ Ti \tag{4-6}$$

In the cases where the product consisted of three or more phases, the amount of each phase could not be calculated based only on elemental analysis.

A series of preliminary experiments were carried out, and examples of the X-ray patterns of the solid products are given in Figure 4.2. These results gave a qualitative idea about the type of intermetallic and elemental phases to be expected at certain experimental conditions. The variation of phase composition observed from X-ray analysis of the product particle is presented in Table 4-1, together with the equilibrium vapor pressures of the reactant chlorides calculated from the vapor pressure temperature relationship [Kubaschewski and Alcock, 1979]. The vapor pressure ratio of $AlCl_3/TiCl_3$ was varied from 4.1 to 112. At the lower range, the product consisted of TiAl and elemental Ti, whereas it contained $TiAl_3$ and elemental Al at higher ratios. The middle range produced a mixture of Ti, TiAl, $TiAl_3$, and Al.

A more quantitative study on the effect of the $AlCl_3/TiCl_3$ ratio was the conducted. Figure 4.3 shows the variation of the mole percent of each phase in the solid product against the input ratio of $AlCl_3/TiCl_3$ at constant partial pressures of Mg and $TiCl_3$ of 3.34 and 1.76 kPa, respectively. The total gas flow rate was constant at 3.0 L/min at 25°C and 86.1 kPa, and the reaction temperature was 920°C. Three distinct regions of product phases were found when the $AlCl_3/TiCl_3$ ratio was varied from 3 to 90.

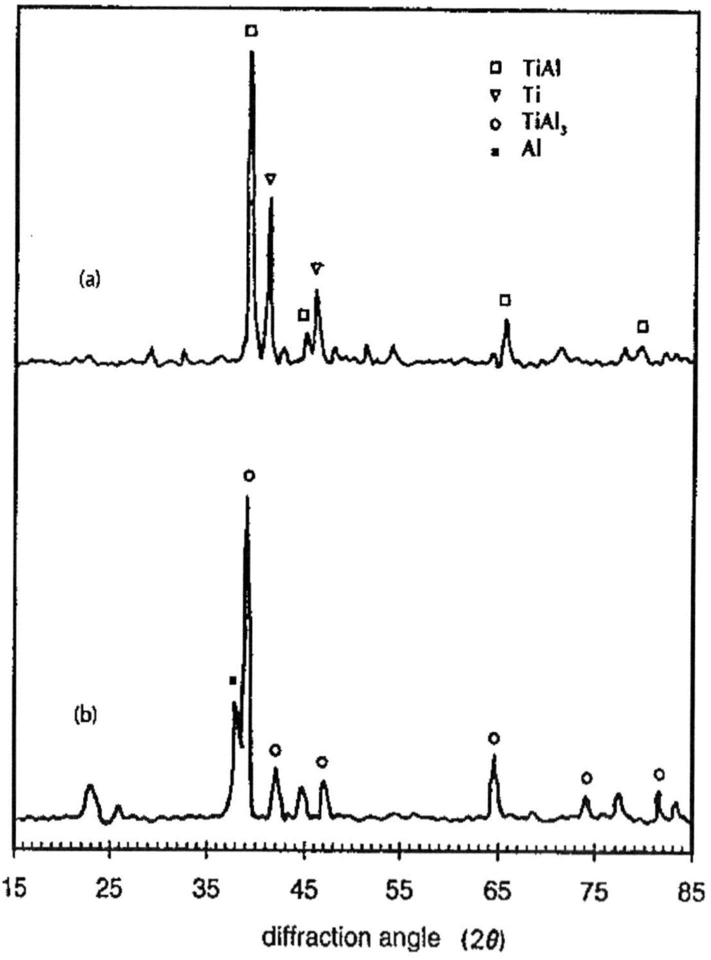

Figure 4.2. X-ray patterns of the solid product (p_{TiCl3} = 0.604 kPa, p_{Mg} = 3.34 kPa, temperature = 920°C, and total flow rate = 3.0 L/min at 25°C and 86.1 kPa). (a) p_{AlCl3} = 4.28 kPa. (*b*) p_{AlCl3}= 9.66 kPa. [Adapted from Sohn and PalDey (1998a)].

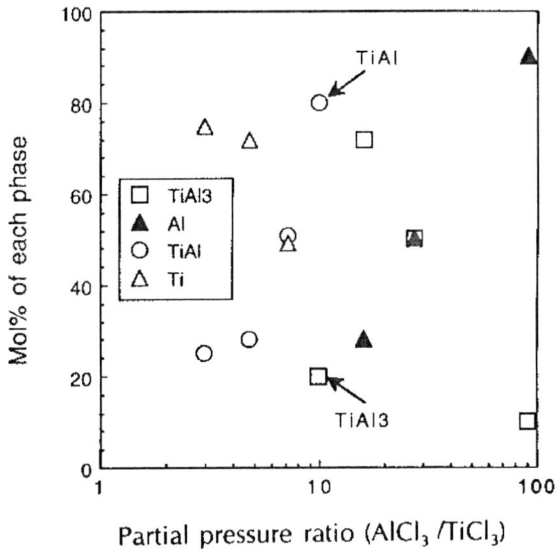

Figure 4.3. Mole percent of each phase in the solid product *vs* $AlCl_3/TiCl_3$ partial pressure ratio. (Experimental conditions are partial pressures of Mg and $TiCl_3$ = 3.34 and 1.76, respectively, temperature = 920°C; and total flow rate = 3.0 L/min at 25°C and 86.1 kPa.) [Adapted from Sohn and PalDey (1998a)].

Table 4.1. Results of the Preliminary Experiments of Magnesium Reduction of $TiCl_3$-$AlCl_3$ Mixtures

Vaporizer Temperature		Equilibrium Vapor Pressure of $TiCl_3$ (kPa)	Ratio of Vapor Pressures $AlCl_3/TiCl_3$	XRD Observation
$TiCl_3$ (°C)	$AlCl_3$ (°C)			
700	200	7.34	4.1	Weak peaks of TiAl, strong peaks of Ti
646	230	2.32	19.4	Comparable peak intensities of TiAl and Ti
600	200	0.77	39.1	Peaks of TiAl, Ti, and $TiAl_3$
646	250	2.32	37.1	Peaks of $TiAl_3$ and TiAl
573	200	0.38	79.26	Peaks of $TiAl_3$, Ti, and Al
600	250	0.77	111.8	Strong peaks of $TiAl_3$; weak peaks of Al

The product consisted of TiAl and Ti between the ratios of 3 to 7. In the ratio range of 7 - 15.5, the product was a mixture of TiAl, $TiAl_3$, and elemental Ti or Al, except at the ratio of 10, where the product consisted of 80 mol pct of TiAl and 20 mol pct of $TiAl_3$. This particular result is indicated by the arrows in Figure 4.3. When the ratio was increased to 16, $TiAl_3$ formed along with elemental Al. The maximum amount of $TiAl_3$ formed was 72 mol pct at this ratio. A further increase in the $AlCl_3/TiCl_3$ ratio favored elemental Al over $TiAl_3$. Ti_3Al did not form within the range of the experimental conditions tested. From the XRD patterns of the solid product, no intermetallic phases between Ti and Mg or Al and Mg were detected.

The experimental results showed a similar trend with those from the thermodynamic calculation, shown in Figure 2.2(a). The equilibrium composition of the solid as a function of the input $AlCl_3/TiCl_3$ ratio indicated three separate regions, as was observed from the experimental results, except that the corresponding region of product composition occurred at considerately different values of the reactant ratios. This might be due to kinetic effects and/or the uncertainties in thermodynamic quantities. The formation of Ti_3Al was predicted by the equilibrium analysis; however, it did not form in the experiments. In some of the experimental conditions, three solid phases formed, indicating that experimental conditions were not at equilibrium, because the formation of three condensed phases is not permitted at equilibrium according to the Gibbs phase rule.

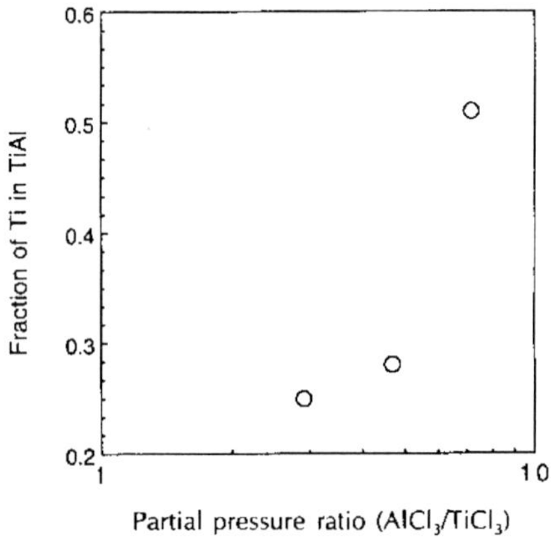

Figure 4.4. Fraction of reduced Ti in TiAl *vs* $AlCl_3/TiCl_3$ partial pressure ratio (the remainder is elemental Ti). [Adapted from Sohn and PalDey (1998a)].

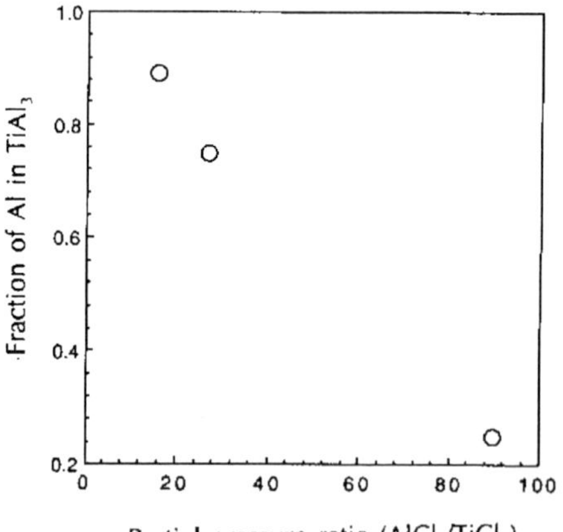

Figure 4.5. Fraction of reduced Al in $TiAl_3$ *vs* $AlCl_3/TiCl_3$ partial pressure ratio (the remainder is elemental Al). [Adapted from Sohn and PalDey (1998a)].

Table 4.2. Conversion of $TiCl_3$ and $AlCl_3$ (Partial Pressure of Mg = 3.34 kPa and Temperature 920°C)

Partial Pressures		Conversion of $TiCl_3$ (Pct)	Conversion of $AlCl_3$ (Pct)
$TiCl_3$ (kPa)	$AlCl_3$ (kPa)		
0.6	1.76	94	8
0.6	2.83	56	3.3
0.4	2.83	50	6.6
0.18	1.76	90	11
0.18	2.83	42	12
0.07	5.94	40	17

Figure 4.4 shows the variation of the fraction of reduced Ti contained in TiAl with the $AlCl_3/TiCl_3$ partial pressure ratio. The fraction of reduced Ti in TiAl increased from 0.25 to 0.51 when the ratio increased from 3 to 7, the rest of the Ti being present as elemental Ti. Figure 4.5 shows the variation of the fraction of reduced Al in the intermetallic product. Within the range of the $AlCl_3/TiCl_3$ ratio from 16 to 90, the maximum fraction of reduced Al contained in $TiAl_3$ was 0.9. The rest of the Al was present as elemental Al. At the ratios between 7 and 16, the product consisted of TiAl, TiA13, and Al.

The conversion degrees of $TiCl_3$ and $AlCl_3$ as functions of their partial pressures are presented in Table 4-2. The maximum conversion degrees were 94 and 17 pct, respectively. TEM observations indicated particles in the size range 200 to 300 nm and the electron diffraction patterns further confirmed the presence of Ti, TiAl, and $TiAl_3$ phases.

To summarize, the co-reduction of $TiCl_3$ and $AlCl_3$ vapor mixtures by magnesium vapor produced $TiAl_3$ and TiAl compounds along with elemental Al or Ti. A pure intermetallic phase did not form, but by carefully choosing the $AlCl_3/TiCl_3$ ratio, a single dominant intermetallic phase along with a small amount of an elemental or another intermetallic phase was obtained. Examples include (a) the formation of 80 mol pct TiAl along with 20 mol pct of $TiAl_3$ at an $AlCl_3/TiCl_3$ partial pressure ratio of 10, Mg and $TiCl_3$ partial pressures of 3.34 and 1.76 kPa, respectively, and temperature of 920°C, and (b) the formation of 72 mol pct $TiAl_3$ with the rest being Al at $AlCl_3/TiCl_3$ = 16 and other conditions being the same as those in (a).

The main advantages of the process are that it is a low-temperature process (titanium aluminides formed at 920°C), because it takes the advantages of the large free energy and exothermic heat of reaction associated with the formation of most intermetallics and the product particles are fine particles in the size range of 200 to 300 nm. The main problem was that a single intermetallic phase did not form; instead, a mixture of two or more phases was produced. This was also predicted by the thermodynamic calculation, which showed that single intermetallic phases formed only at very narrow ranges of the reactant concentrations. By carefully choosing the reactant partial pressures, the content of a desired intermetallic phase was maximized in the experiments. The formation of a mixture of two intermetallic phases may be desirable in some cases. As was mentioned in the introduction, a dual-phase (TiAl + Ti_3Al) alloy is used in some gas turbine rotational components. In the experimental work discussed here, a dual-phase mixture of 80 mol pct TiAl + 20 mol pct $TiAl_3$ was produced.

$TiCl_3$ was used in this work rather than $TiCl_4$, which is the conventional intermediate in titanium production, for reasons given above. The findings of this investigation, however, are expected to be largely applicable even when $TiCl_4$ is used as would be the case in a large-scale production based on this process.

4.2. Nickel Aluminides by Magnesium Reduction

The chemical vapor synthesis (CVS) process was applied to the synthesis of nickel aluminide particles by Sohn and PalDey [1998b]. NiAl, $NiAl_3$, and Ni_2Al_3 were formed by reducing mixtures of $NiCl_2$ and $AlCl_3$ vapors. The effects of $AlCl_3/NiCl_2$ and $Mg/NiCl_2$ partial pressure ratios on the formation of nickel aluminides were studied. The maximum NiAl content obtained was 98 mol pct. At the $AlCl_3/NiCl_2$ ratio of 19, a two-phase mixture of 17 mol pct of NiAl + 83 mol pct of $NiAl_3$ was produced. The product particles were very fine in the size range of 100 to 200 nm.

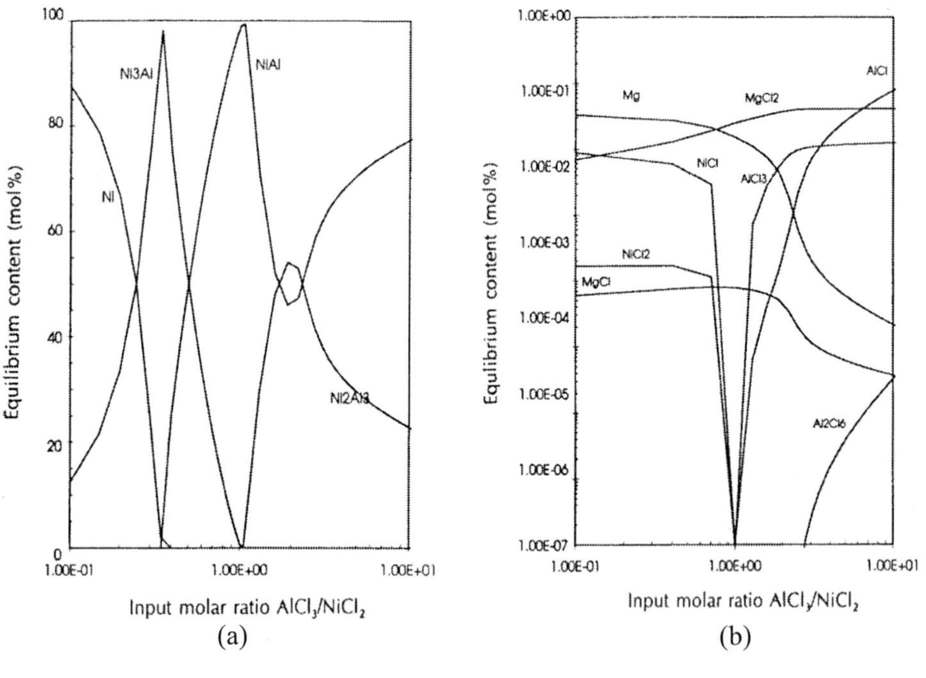

Figure 4.6. (*a*) Calculated effect of the input molar ratio $AlCl_3/NiCl_2$ on the equilibrium solid phase composition in the Ni-Al-Mg-Cl-Ar system (input molar ratio $Mg/NiCl_2 = 4.0$, total pressure = 101.3 kPa, and temperature = 1000°C). (*b*) Calculated effect of the input molar ratio $AlCl_3/NiCl_2$ on the equilibrium gas phase composition in the Ni-Al-Mg-Cl-Ar system (input molar ratio $Mg/NiCl_2 = 4.0$, total pressure = 101.3 kPa, and temperature = 1000°C). [Adapted from Sohn and PalDey (1998b)].

Thermodynamic Equilibrium Analysis

Sohn and PalDey [1998b] first carried out equilibrium calculation for the Ni-Al-Mg-Cl-Ar system using the aforementioned HSC software [Outotec, 2002]. Equilibrium compositions of the solid elemental and intermetallic phases were calculated as functions of reactant concentration and temperature.

Figure 4.6(a) shows the effect of the input $AlCl_3/NiCl_2$ ratio on the equilibrium contents of the solid phases at a constant $Mg/NiCl_2$ ratio of 4.0. In the range of the $AlCl_3/NiCl_2$ ratio between 0.1 and 10, the contents of Ni_3Al and NiAl reach a maximum value of 99.5 mol pct at the $AlCl_3/NiCl_2$ ratios of 0.33 and 1.0, respectively. Beyond the ratio of 1.0, Ni_2Al_3 starts forming at the expense of NiAl and reaches a maximum value of 53 mol pct at the $AlCl_3/NiCl_2$ ratio 1.5. The Ni_2Al_3 content decreases and NiAl becomes more favorable beyond this point. The corresponding equilibrium gas-phase composition shown in Figure 4.6(b) indicates that beyond the ratio of 1.0, the gas phase contains a large excess of $AlCl_3$ and AlCl. The amount of Mg decreases to a negligible value beyond the $AlCl_3/NiCl_2$ ratio of 2.0, and thus Mg is not available to reduce the excess $AlCl_3$ to form elemental Al.

Figure 4.7 presents the effect of the input $Mg/NiCl_2$ partial pressure ratio on the equilibrium solid-phase composition at a constant $AlCl_3/NiCl_2$ ratio of 1.0. The composition changes from Ni and Ni_3Al in the lower range of the ratio to Ni_3Al and NiAl in the higher range as the $Mg/NiCl_2$ ratio increases from 1 to 10. Ni_2Al_3 or elemental Al does not form at the $AlCl_3/NiCl_2$ ratio of 1.0.

The effect of temperature in the range of 900°C to 1300°C on the equilibrium solid-phase composition is shown in Figure 4.8. It is seen that a lower temperature favors the formation of Ni_2Al_3. The NiAl content goes through a maximum value at 1200°C.

To summarize, the theoretical equilibrium analysis of the Ni-Al-Mg-Cl-Ar system at different reactant partial pressures and temperatures predicts the formation of several intermetallic phases along with some individual metals. A single intermetallic phase was possible only at a very narrow range of the partial pressure ratio of reactants; for all other conditions, the solid product was predicted to be a mixture of two phases. The contents of Ni_3Al and NiAl reached essentially 100 mol pct at the $AlCl_3/NiCl_2$ ratios 0.33 and 1.0, respectively. The maximum Ni_2Al_3 content was 53 pct mol pct at an $AlCl_3/NiCl_2$ ratio of 1.5.

Experimental Work

The experimental apparatus and procedure presented in detail in the previous section on the synthesis of titanium aluminides were also used in this work by Sohn and PalDey [1998b]. The effect of reactant partial pressures on the formation of different intermetallic phases was investigated experimentally. The size and shape of the fine particles were observed by transmission electron microscopy (TEM). In the experiments thus carried out, NiAl, Ni_2Al_3, and $NiAl_3$ intermetallic compounds were formed, along with elemental Ni and Al, by the magnesium vapor reduction of mixed nickel and aluminum chlorides. The governing chemical reactions are as follows:

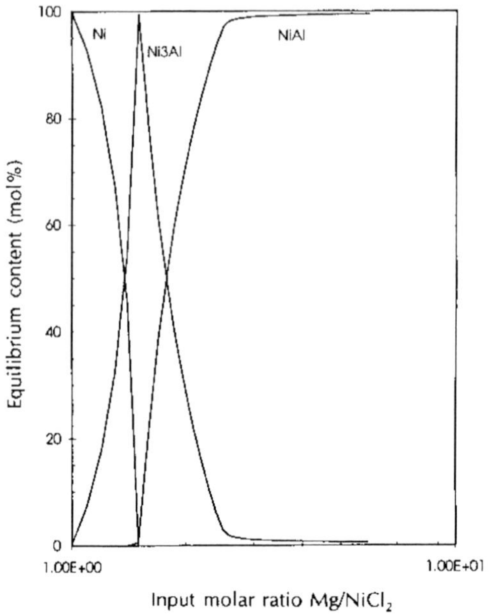

Figure 4.7. Calculated effect of the input molar ratio $Mg/NiCl_2$ on the equilibrium solid phase composition in the Ni-Al-Mg-Cl-Ar system (input molar ratio $AlCl_3/NiCl_2$ = 1.0, total pressure = 101.3 kPa, and temperature = 1000°C). [Adapted from Sohn and PalDey (1998b)].

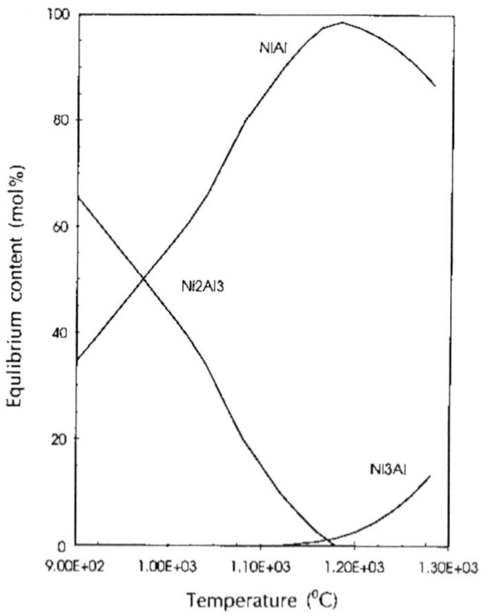

Figure 4.8. Calculated effect of temperature on the equilibrium solid phase composition in the Ni-Al-Mg-Cl system (input molar ratio $AlCl_3/NiCl_2$ = 1.5, input molar ratio $Mg/NiCl_2$ = 4.0, and total pressure 101.3 kPa). [Adapted from Sohn and PalDey (1998b)].

$$NiCl_2(g) + AlCl_3(g) + 2.5Mg(g) = NiAl(s) + 2.5MgCl_2(l,g) \quad (4\text{-}7)$$

$$2NiCl_2(g) + 3AlCl_3(g) + 6.5Mg(g) = Ni_2Al_3(s) + 6.5MgCl_2(l,g) \quad (4\text{-}8)$$

$$NiCl_2(g) + 3AlCl_3(g) + 5.5Mg(g) = NiAl_3(s) + 5.5MgCl_2(1,g) \quad (4\text{-}9)$$

$$NiCl_2(g) + Mg(g) = Ni(s) + MgCl_2(1,g) \quad (4\text{-}10)$$

$$AlCl_3(g) + 1.5Mg(g) = Al(1) + 1.5MgCl_2(1,g) \quad (4\text{-}11)$$

These synthesis reactions proceeded without any nucleation promoters, in contrast to the synthesis of nickel aluminides by a similar process but using hydrogen instead of magnesium [Sohn and PalDey, 1998d] which requires the presence of a small amount of aluminum vapor. This will be further discussed in Section 4.3.

Examples of the X-ray patterns of the solid products are given in Figure 4.9. The effect of the input $AlCl_3/NiCl_2$ ratio was first determined. The partial pressures of $NiCl_2$ and Mg were kept constant at 1.26 and 5.4 kPa, respectively. The reaction temperature was 1000°C, and the total input flow rate was 3 L/min at 25°C and 86.1 kPa. The solid-phase composition changed from NiAl + Ni to NiAl + Al and then to NiAl + $NiAl_3$, as the $AlCl_3/NiCl_2$ ratio was increased from 0.64 to 19, as shown in Figure 4.10. The solid phase consisted of 81 mol pct NiAl, the rest being elemental Ni, at the $AlCl_3/NiCl_2$ ratio of 0.64, and it consisted of NiAl and Al at a higher ratio. The maximum NiAl content obtained was 98 mol pct. At a much higher $AlCl_3/NiCl_2$ ratio value of 19, the product consisted of 17 mol pct of NiAl and 83 mol pct of $NiAl_3$. No intermetallic compounds between Ni and Mg or Al and Mg were detected.

The variation of the fraction of reduced Al in NiAl and $NiAl_3$ with $AlCl_3/NiCl_2$ ratio is presented in Figure 4.11, showing an increase from 0.94 to 0.98 as the ratio varied from 1.07 to 2.14. The reduced Al was distributed between NiAl and $NiAl_3$, as shown by arrows in the figure at an $AlCl_3/NiCl_2$ ratio of 19.

Figure 4.12 shows the effect of input $Mg/NiCl_2$ partial pressure ratio on the composition of the solid product phases at the $NiCl_2$ and $AlCl_3$ partial pressures of 1.26 and 2.14 kPa, respectively. The reaction temperature was 1000°C, and the total input flow rate was 3 L/min at 25°C and 86.1 kPa. At the $Mg/NiCl_2$ ratio of 2.1, the product contained 70 mol pct of NiAl, the rest being elemental Ni. The NiAl content increased to 98 mol pct, the rest being elemental Al instead of Ni, with a further increase of the ratio to 4.3. At a higher $Mg/NiCl_2$ ratio of 8.6, the solid product contained 32 mol pct Ni_2Al_3 and 68 mol pct Al.

The conversion degrees of $NiCl_2$ and $AlCl_3$ as functions of their partial pressures are presented in Table 4-3. The maximum conversion degrees of $NiCl_3$ and $AlCl_3$ were 78 and 35 pct at the partial pressures 0.34 and 1.35 kPa, respectively. TEM observations revealed chains of spherical particles in the size range 100 to 200 nm and the electron diffraction patterns identified NiAl, $NiAl_3$, and Ni phases [Sohn and PalDey, 1998b].

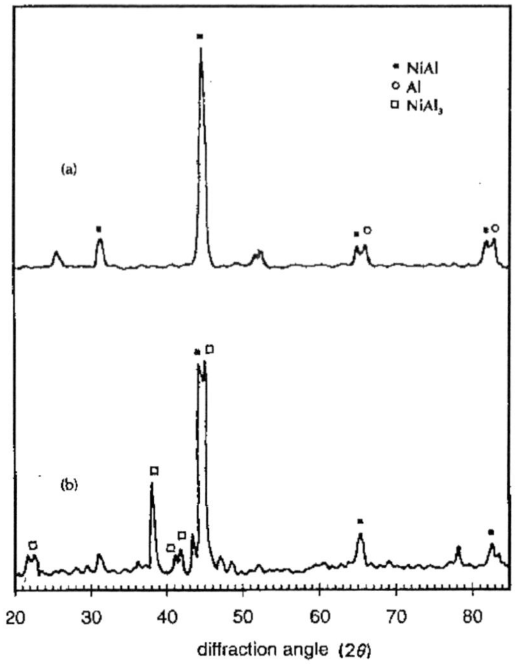

Figure 4.9. X-ray patterns of the solid product (p_{NiCl2} = 1.26 kPa, p_{Mg} = 5.5 kPa, temperature = 920°C, and total flow rate = 3.0 L/min at 25°C and 86.1 kPa). (*a*) p_{AlCl3} = 1.38 kPa. (*b*) p_{AlCl3} = 24 kPa. [Adapted from Sohn and PalDey (1998b)].

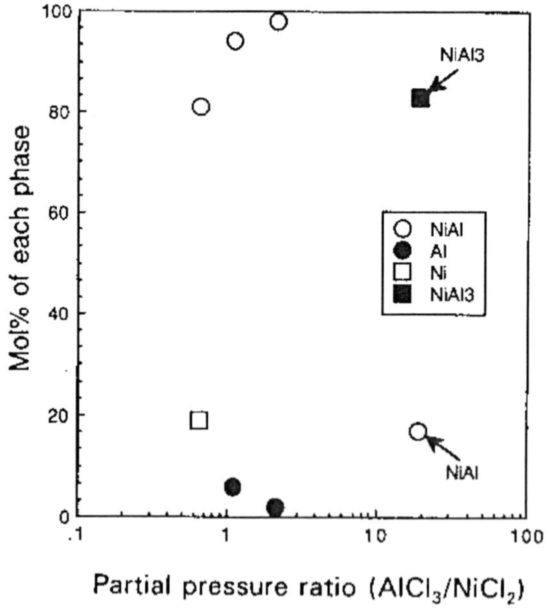

Figure 4.10. Mole percent of each phase in the solid product *vs* the input $AlCl_3/NiCl_2$ partial pressure ratio. (Experimental conditions are partial pressures of Mg and $NiCl_2$ = 5.41 and 1.26 kPa, respectively; temperature = 1000°C; and total flow rate 3 L/min at 25°C and 86.1 kPa.) [Adapted from Sohn and PalDey (1998b)].

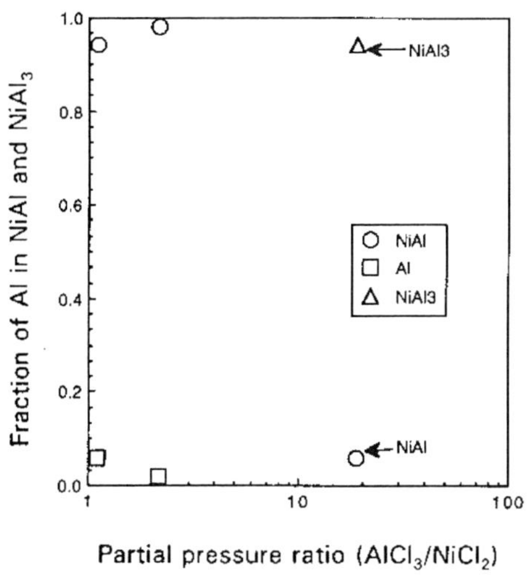

Figure 4.11. Fraction of reduced Al in NiAl and $NiAl_3$ *vs* partial pressure ratio. [Adapted from Sohn and PalDey (1998b)].

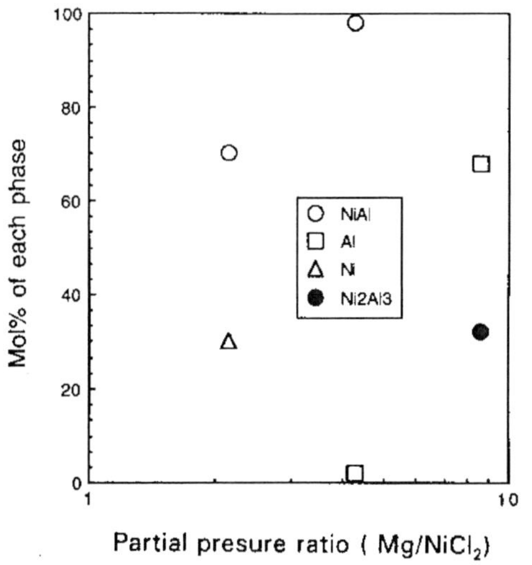

Figure 4.12. Mole percent of each phase in the solid product *vs* the input $Mg/NiCl_2$ partial pressure ratio. (Experimental conditions are partial pressures of $AlCl_3$ and $NiCl_2$ = 2.14 and 1.26 kPa, respectively; temperature = 1000°C; and total flow rate 3 L/min at 25°C and 86.1 kPa.) [Adapted from Sohn and PalDey (1998b)].

Table 4.3. Conversion of $NiCl_2$ and $AlCl_3$ (Partial Pressure of Mg = 5.41 kPa, Temperature 1000°C, and Total Flow Rate 3 L/min at 25°C and 86.1 kPa)

Partial Pressures		Conversion of $TiCl_3$	Conversion of $AlCl_3$
$NiCl_2$ (kPa)	$AlCl_3$ (kPa)	(Pct)	(Pct)
0.34	6.94	78	4.2
1.26	1.35	38	35
2.61	1.35	33	30
1.26	2.7	32	17

The thermodynamic equilibrium calculation predicted the formation of Ni_3Al, NiAl, and Ni_2Al_3 along with elemental Ni and Al in the range of the experimental variables tested. Instead, NiAl, $NiAl_3$, and Ni_2Al_3 were produced experimentally. The variation of input $AlCl_3/NiCl_2$ ratio at a constant $Mg/NiCl_2$ ratio of 5.41 produced mixtures of NiAl and $NiAl_3$ along with elemental Ni or Al, as can be seen in Figure 4.10. In the thermodynamic equilibrium analysis, shown in Figure 4.6(a), Ni_3Al, NiAl, and Ni_2Al_3, but not $NiAl_3$, were predicted as the stable solid phases. This discrepancy indicates that experimental conditions might not have been at equilibrium. The product consisted of NiAl along with elemental Ni or Al in the $Mg/NiCl_2$ range of 2.14 to 4.28, as shown in Figure 4.12. At a higher $Mg/NiCl_2$ ratio value of 8.57, the product contained Ni_2Al_3 and elemental Al. In the equilibrium analysis shown in Figure 4.7, Ni, Ni_3Al, and NiAl were indicated.

To summarize, the magnesium vapor reduction of mixtures of $AlCl_3$ and $NiCl_2$ vapors produced NiAl, Ni_2Al_3, and $NiAl_3$ intermetallic compounds. The contents of the desired intermetallic compounds were maximized by adjusting the $AlCl_3/NiCl_2$ and $Mg/NiCl_2$ partial pressure ratios. At 1000°C, the NiAl content was maximized to 98 mol pct at $AlCl_3$, $NiCl_2$, and Mg partial pressures of 2.7, 1.26, and 5.4 kPa, respectively. A mixture of 17 mol pct of NiAl + 83 mol pct of $NiAl_3$ was formed at the partial pressures of $AlCl_3$, $NiCl_2$, and Mg of 6.94, 1.26, and 5.4 kPa, respectively. The product particle size was in the range of 100 to 200 nm, as observed by TEM.

4.3. Nickel Aluminides by Hydrogen Reduction

The chemical vapor synthesis (CVS) reactions for preparing intermetallic powders by hydrogen reduction can in general be written as follows:

$$mMCl_x(g) + nNCl_y(g) + 0.5(mx + ny)H_2 = M_mN_n(s) + (mn + ny)HCl(g) \quad (4\text{-}12)$$

where M and N represent two different metals, with x and y being the valences, and M_mN_n the intermetallic compound formed.

Sohn and PalDey [1998d] synthesized for the first time nickel aluminide (Ni_3Al) particles (50-100 nm) at 900 to 1150°C using hydrogen as the reducing agent according to the above reaction equation. The fact that aluminum chloride is reduced by this reaction scheme is very significant, because the reduction of $AlCl_3$ alone by hydrogen is thermodynamically unfavorable at these moderate temperatures. However, the negative free energy of formation of the intermetallic compound makes the overall reaction feasible, as can be seen from the following equations:

$$AlCl_3(g) + 1.5H_2 = Al(l) + 3HCl(g); \quad (4\text{-}13)$$

$\Delta G° = +352$ kJ at 1000°C

$$3NiCl_2(g) + AlCl_3(g) + 4.5H_2 = Ni_3Al(s) + 9HCl(g); \quad (4\text{-}14)$$

$\Delta G° = -160$ kJ at 1000°C

More detailed thermodynamic equilibrium calculations for the Ni–Al–H–Cl–Ar system were first carried out to evaluate the effect of the reactant partial pressures and temperature on the formation of intermetallic phases. A single intermetallic phase was found to be feasible only in a very narrow range of the reactant partial pressures. For all other conditions the predicted solid product was a mixture of two phases. Experimentally, Ni_3Al was formed along with elemental Ni. Although the formation of Ni_3Al by the co-reduction of $NiCl_2$ and $AlCl_3$ by H_2 at 1100°C is thermodynamically feasible, it did not occur experimentally in this work. However, the co-reduction reaction proceeded as predicted by thermodynamics when a low vapor pressure of aluminum was provided. The effects of reactant partial pressures and temperature were studied. The Ni_3Al content was maximized to 52 mol% at 1050°C under the H_2, $AlCl_3$, and $NiCl_2$ partial pressures of 57, 1.5, and 0.5 kPa, respectively. The product particles, as observed by TEM, were very fine, but usually agglomerated. The electron diffraction analysis identified the particles of NiAl and $NiAl_3$ along with Ni_3Al and elemental Ni.

Thermodynamic Equilibrium Analysis

Sohn and PalDey [1998d] carried out equilibrium calculations for the Ni–Al–H–Cl–Ar system by the use of HSC Chemistry software [Outotec, 2002]. Figure 4.13 shows the effect of $H_2/NiCl_2$ molar ratio on the equilibrium contents of the solid phases at 1000°C. As the ratio varies from 10 to 10,000, the mol% of Ni_3Al increases from a negligible value to a maximum of 99 mol% at the $H_2/NiCl_2$ ratio of 300, and then its content decreases while the formation of NiAl becomes more favorable. The effect of $AlCl_3/NiCl_2$ molar ratio, shown in Figure 4.14(a), is similar to that of $H_2/NiCl_2$ ratio.

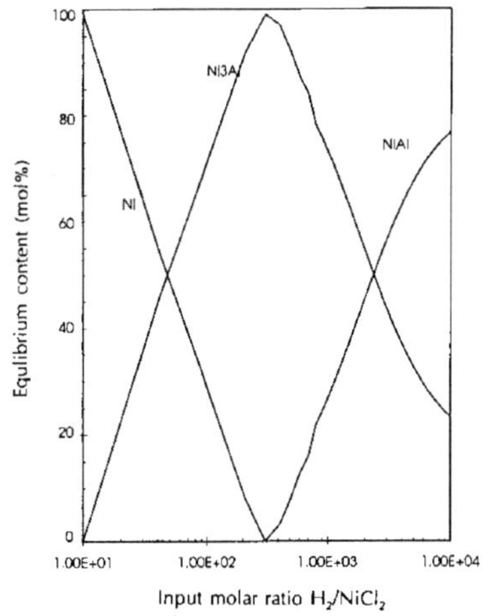

Figure 4.13. Calculated effect of the input molar ratio $H_2/NiCl_2$ on the equilibrium solid-phase composition in the Ni–Al–H–Cl–Ar system. (Input molar ratio $AlCl_3/NiCl_2 = 1.0$; total pressure = 101.3 kPa; temperature = 1000°C.) [Adapted from Sohn and PalDey (1998d)].

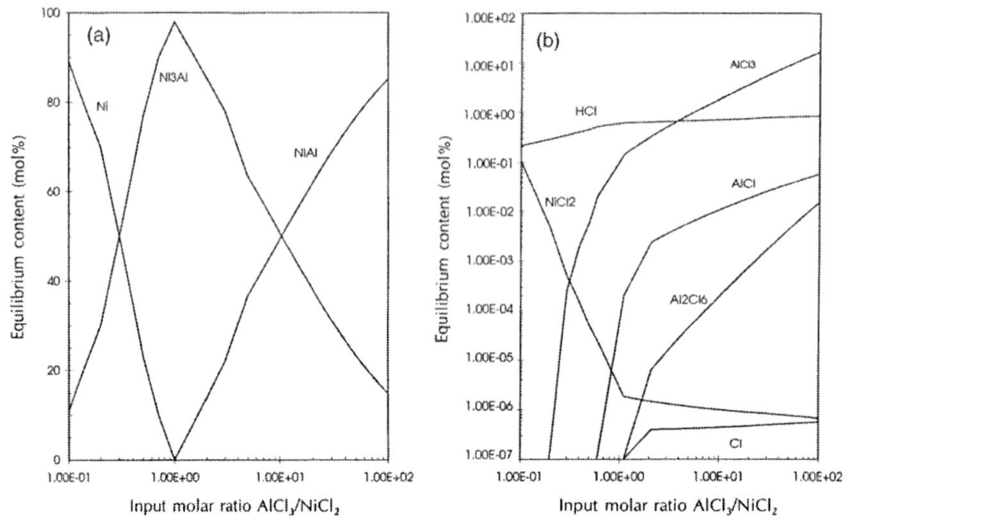

(a) (b)

Figure 4.14. (a) Calculated effect of the input molar ratio $AlCl_3/NiCl_2$ on the equilibrium solid-phase composition in the Ni–Al–H–Cl–Ar system. (Input molar ratio $H_2/NiCl_2$ = 500; total pressure = 101.3 kPa; temperature = 1000°C.) (b) Calculated effect of the input molar ratio $AlCl_3/NiCl_2$ on the equilibrium gas-phase composition in the Ni–Al–H–Cl–Ar system. (Input molar ratio $H_2/NiCl_2$ = 500; total pressure = 101.3 kPa; temperature = 1000°C.) [Adapted from Sohn and PalDey (1998d)].

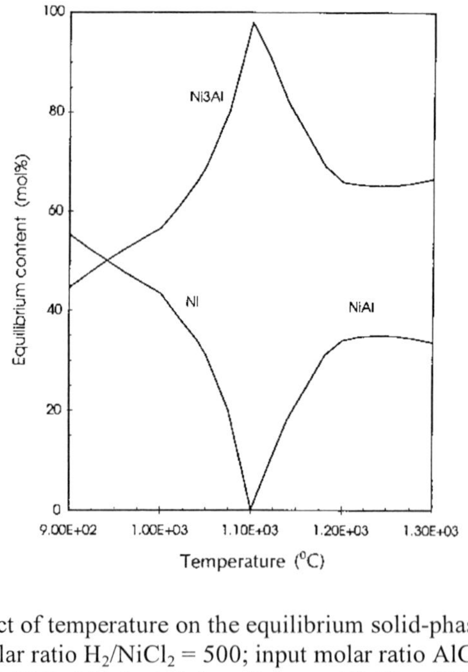

Figure 4.15. Calculated effect of temperature on the equilibrium solid-phase composition in the Ni–Al–H–Cl–Ar system. (Input molar ratio $H_2/NiCl_2$ = 500; input molar ratio $AlCl_3/NiCl_2$ = 1.0; total pressure = 101.3 kPa.) [Adapted from Sohn and PalDey (1998d)].

Figures 4-13 and 4-14(a) indicate that higher ratios of $H_2/NiCl_2$ and $AlCl_3/NiCl_2$ favor the formation of NiAl, whereas Ni_3Al goes through a maximum value at $H_2/NiCl_2$ = 300 and $AlCl_3/NiCl_2$ = 1.0. The effect of temperature on the equilibrium composition is shown in Figure 4.15. NiAl formation is predicted only at temperatures higher than 1100°C, replacing the formation of elemental Ni at lower temperatures. The formation of Ni_3Al, which occurs in the entire temperature range used, goes through a maximum of 99 mol% at 1100°C. A pure Ni_3Al intermetallic phase forms only at very narrow ranges of the partial pressures of reactants. For all other conditions, the solid is predicted to be a mixture of two phases. However, by optimizing the reaction conditions, the content of a particular desirable intermetallic phase can be maximized. As an example, higher partial pressure ratios of $H_2/NiCl_2$ and $AlCl_3/NiCl_2$ maximize the NiAl content to 85 mol %. An increase of total pressure above 1 atm had no significant effects on the equilibrium composition.

Experimental Work

Sohn and PalDeỳ [1998d] carried out experiments for the preparation of intermetallic compounds in the same quartz reactor as described in the previous section. Although the formation of Ni_3Al is thermodynamically feasible at 1000°C, as shown by Eq. (4-14) as well as in Figuews 4-13 through 4-15, it did not take place easily. Wide ranges of reactant partial pressures and temperature were tried. The ranges of $NiCl_2$, $AlCl_3$, and H_2 partial pressures were 0.41 - 1.11, 0.6 - 4.2, and 11.6 - 61.3 kPa, respectively. Temperature was varied from 950 to 1150°C. Instead of intermetallic phases, only elemental Ni was obtained.

It was reasoned that the system needed a promoter to overcome the barrier of the thermodynamically unfavorable hydrogen reduction of $AlCl_3$. It was thought that elemental aluminum would be the natural choice. When a small amount of aluminum vapor was added to the $NiCl_2$ and $AlCl_3$ vapor mixture, the Ni_3Al intermetallic phase was indeed formed. An aluminum melt at 1100°C in an alumina boat of 1 x 3 cm size placed inside the reaction tube (8 in Figure 4.1 shown in the previous section) acted as the source of the Al vapor. $NiCl_2$ and $AlCl_3$ were vaporized in the respective vaporizers (5, 6 in Figure 4.1) and were carried by argon gas into the reaction zone. Here, the co-reduction of $NiCl_2$ and $AlCl_3$ by H_2 took place at a very low aluminum vapor pressure of 2.3×10^{-6} atm, according to reaction (4-14). Three different sets of experiments were run in order to test whether $AlCl_3$ was reduced to form the aluminide or only the Al vapor reacted with $NiCl_2$ in the presence of H_2. Table 4-4 shows the conditions of the different experiments and the corresponding observations from XRD and chemical analysis. In the first set of experiments with $NiCl_2$ and $AlCl_3$, only Ni peaks were observed in the solid product phases. In the second set of experiments with $NiCl_2$, $AlCl_3$, and aluminum vapor at a very low vapor pressure in the order of 10^{-6} atm, strong peaks of Ni_3Al along with Ni peaks were observed. No aluminide peak was observed in the third set of experiments with only $NiCl_2$ and aluminum vapor with no $AlCl_3$. However, when the same experiment was done (experiment 4 in Table 4-4) at a much lower total gas flow rate and thus a longer length of reaction time, weak peaks of aluminide along with strong peaks of Ni were observed in the XRD. Hydrogen gas flow rate, reaction temperature, and total gas flow rate were constant in the first three cases.

The amount of the nickel aluminide formed in the second case was much larger than that formed in the third or the fourth case. Thus, it can be concluded that, in the second case, the co-reduction of $NiCl_2$ and $AlCl_3$ to form Ni_3Al took place. Aluminum vapor, though present at a very low partial pressure, promoted the reduction of $AlCl_3$ by forming fine nuclei of the aluminide.

Table 4.4. Tests for the co-reduction of $AlCl_3$ and $NiCl_2$

Experiments	XRD analysis	Elemental analysis
1. With $NiCl_2$ and $AlCl_3$	Ni peaks only; no aluminide peaks	No Al
2. With $NiCl_2$, $AlCl_3$, and Al vapor	Strong Ni_3Al peaks along with Ni peaks	Al
$NiCl_2$ and Al vapor with no $AlCl_3$	Ni peaks only	Al undetected
Same as case 3 but at a much lower total flow rate of gas	Weak peaks of aluminide along with Ni peaks	Some Al

The requirement of the presence of aluminum vapor for the reaction to proceed is in contrast to the synthesis of Ni_4Mo by a similar process, discussed in the next section; that reaction proceeded without any such nucleation promoters [Sohn and PalDey, 1998c]. Further, when magnesium rather than hydrogen is used as the reducing agent, the $NiCl_2$ + $AlCl_3$ mixture is reduced to produce nickel aluminides without any promoters [Sohn and PalDey, 1998b], as discussed in the previous section. In these two cases, the reducing agents can reduce the component metal chlorides individually in similar conditions and the intermetallic phase formation further decreases the free energy.

The effects of different experimental variables on the formation of nickel aluminides by hydrogen reduction will now be discussed.

Effect of $AlCl_3/NiCl_2$: Figure 4.16 shows the variation of the Ni_3Al content against the $AlCl_3/NiCl_2$ partial pressure ratio at constant H_2 and $NiCl_2$ partial pressures of 33.6 and 0.55 kPa, respectively. The reaction temperature was 1100°C and the total flow rate was kept constant at 4.5 L/min at 25°C and 86.1 kPa. The Ni_3Al content increased from 18 to 35 mol% when the $AlCl_3/NiCl_2$ ratio increased from 1.0 to 6.7. The rest of the product was elemental Ni. A further increase in the $AlCl_3/NiCl_2$ ratio produced a mixture of three phases and it was not possible to calculate the amount of each phase present.

Figures 4-17(a) and 4-17(b) show the XRD of product phases at $AlCl_3/NiCl_2$ ratios of 12 and 18, respectively, other conditions being the same. Some peaks other than those of Ni and Ni_3Al are observed.

Although these peaks mostly match those of $NiAl_3$ and NiAl compounds, not all the major peaks of NiAl and $NiAl_3$ are present. Thus, it is difficult to conclude whether NiAl and $NiAl_3$ formed. However, the elemental analysis of the product presented in Table 4-5 shows that the wt% of reduced aluminum in the product phase was substantially larger in the cases of the two highest partial pressures of $AlCl_3$ than in all other cases, indicating that the amount of Al contained in the intermetallic phases were greater. The fraction of reduced Ni contained in Ni_3Al against $AlCl_3/NiCl_2$ ratio is shown in Figure 4.18, which was obtained from Figure 4.16. Starting from 0.4, the fraction of reduced Ni in Ni_3Al increased to 0.6 by increasing the $AlCl_3/NiCl_2$ ratio by 7 times. The rest of the Ni was present as elemental Ni.

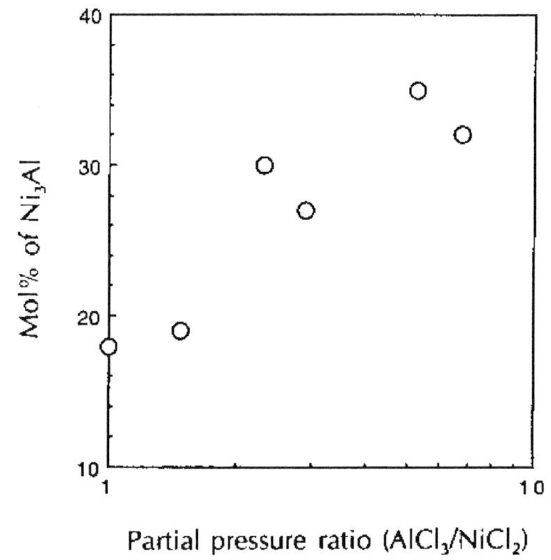

Figure 4.16. Ni_3Al content in the solid product vs input $AlCl_3/NiCl_2$ partial pressure ratio. (Experimental conditions are partial pressure of H_2 and $NiCl_2$ = 33.6 and 0.55 kPa, respectively; temperature = 1100°C and total flow rate = 4.5 L/min at 25°C and 86.1 kPa.) [Adapted from Sohn and PalDey (1998d)].

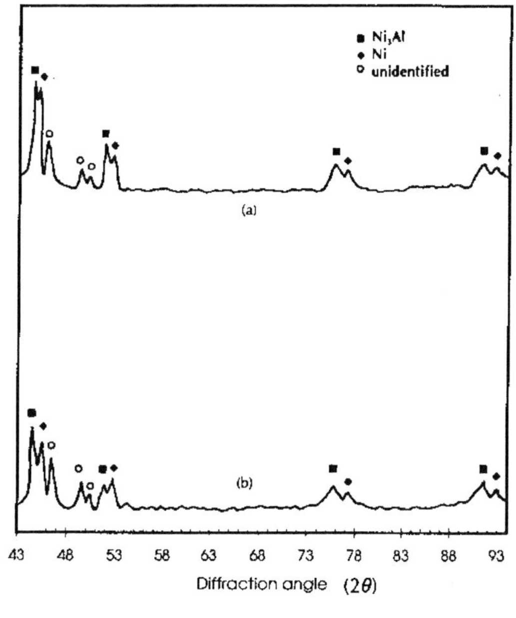

Figure 4.17. X-ray diffraction pattern of the solid product phases in the Ni–Al system at $AlCl_3/NiCl_2$ partial pressure ratio of (a) 12 and (b) 18. (Other experimental conditions are the same as those presented in Figure 5.) [Adapted from Sohn and PalDey (1998d)].

Table 4.5. Analysis of the product phases in the Ni–Al system. (Partial pressure of H_2 = 33.6 kPa, $NiCl_2$ = 0.55 kPa, temperature 1100°C, and total flow rate 4.5 L/min at 25°C and 86.1 kPa.)

p_{AlCl3}/ p_{NiCl2}	Total Ni in product (mol %)	Total Al in product (mol %)	Observation from XRD of product
1.0	88	12	Peaks of Ni_3Al; no Al
1.5	88	12	Peaks of Ni_3Al; no Al
2.3	86	14	Peaks of Ni_3Al; no Al
2.9	84	16	Peaks of Ni_3Al; no Al
6.7	82	18	Peaks of Ni_3Al; no Al
12	77	23	A few unknown peaks along with peaks of Ni_3Al and Ni; no Al
18	72	28	Strong peaks of unknown phase; no Al

Effect of $H_2/NiCl_2$ partial pressure ratio: The variation of Ni_3Al content with the $H_2/NiCl_2$ ratio was investigated at 1100°C and constant $AlCl_3$ and $NiCl_2$ partial pressures of 3.3 and 0.5 kPa, respectively. The product phase consisted of Ni_3Al and Ni. The Ni_3Al content increased from 19 to 46 mol% as the partial pressure ratio was increased from 18 to 100, as can be seen in Figure 4.19. The rest of the product was elemental Ni. No other phases were formed within the range of partial pressure of H_2 tested. The fraction of reduced Ni contained in Ni_3Al was found to increase from 0.41 to 0.72, as can be observed in Figure 4.20.

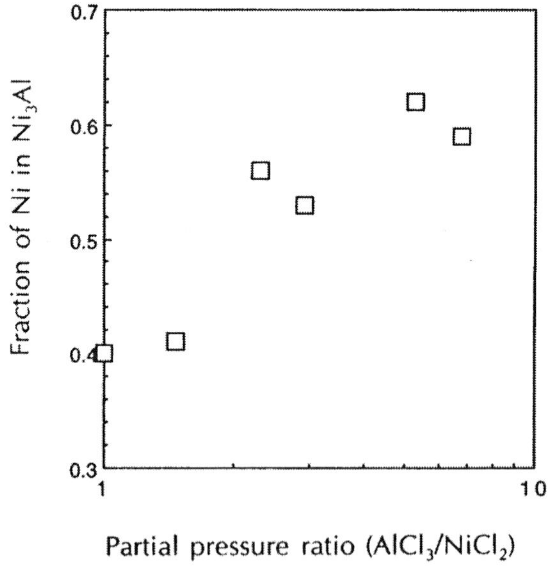

Figure 4.18. The fraction of reduced Ni in Ni_3Al versus input $AlCl_3/NiCl_2$ partial pressure ratio. [Adapted from Sohn and PalDey (1998d)].

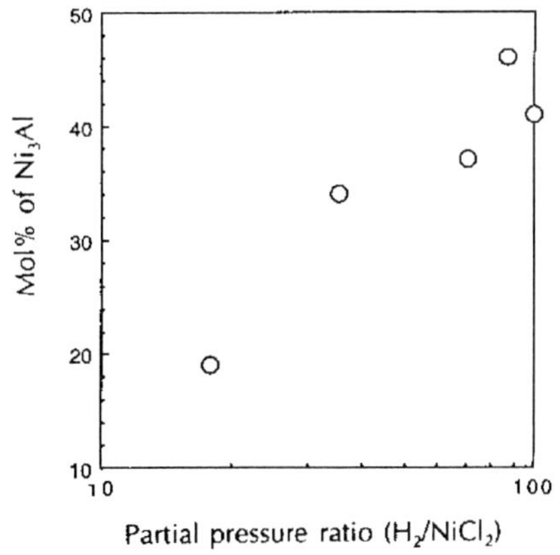

Figure 4.19. Ni_3Al content in the solid product versus input $H_2/NiCl_2$ partial pressure ratio. (Experimental conditions are partial pressure of $AlCl_3$ and $NiCl_2$ = 3.3 and 0.5 kPa, at 25°C at 86.1 kPa.) [Adapted from Sohn and PalDey (1998d)].

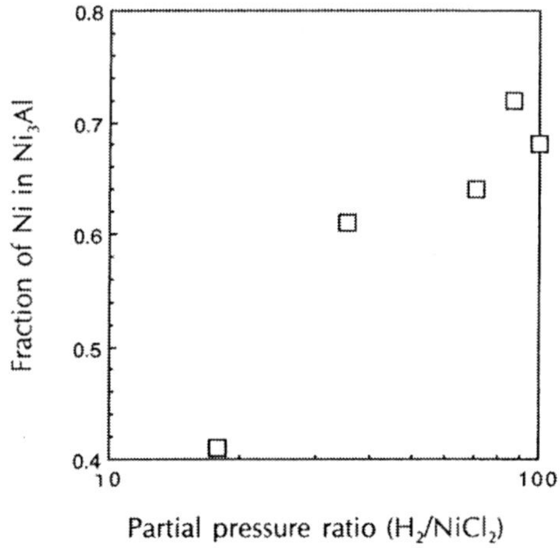

Figure 4.20. The fraction of reduced Ni in Ni_3Al versus input $H_2/NiCl_2$ partial pressure ratio. [Adapted from Sohn and PalDey (1998d)].

Effect of temperature: The effect of temperature on the composition of the solid product was evaluated at constant partial pressures of reactants. The total flow rate was 4.5 L/min at 25°C and 86.1 kPa and the partial pressures of H_2, $AlCl_3$, and $NiCl_2$ were 57, 1.4, and 0.5 kPa, respectively. Starting from 19 mol% at 900°C, the Ni_3Al content reached a maximum of 53 mol% at 1050°C and then decreased with a further increase in temperature, as shown in Figure 4.21. The rest of the product was elemental Ni. The variation in the fraction of reduced Ni contained in Ni_3Al with temperature is shown in Figure 4.22, the maximum value being 0.8 at 1050°C. Comparison between experimental results and the results of equilibrium calculation: The equilibrium calculation predicted the formation of Ni_3Al, NiAl, and elemental Ni in the Ni–Al–H–Cl–Ar system. Experimentally, only Ni_3Al was obtained along with elemental Ni. At higher $AlCl_3/NiCl_2$ partial pressure ratios, the XRD patterns of the product, presented in Figure 4.17, showed a few unidentified peaks that matched with some of the peaks of NiAl and $NiAl_3$. Ni_3Al formed along with Ni in the $AlCl_3/NiCl_2$ range from 1 to 7, as shown in Figure 4.16. The product consisted of Ni_3Al, Ni, and another phase which may be NiAl or $NiAl_3$, at higher ratios of 12 and 18. This observation is consistent with the calculated results presented in Figure 4.14(a), which predicts that beyond the $AlCl_3/NiCl_2$ ratio of 10, the solid product should consist mostly of NiAl with little Ni_3Al. Experimental results presented in Figure 4.19 show that the Ni_3Al content increased continuously as the $H_2/NiCl_2$ ratio increased from 10 to 100. The equilibrium Ni_3Al content also increased in this range of $H_2/NiCl_2$ at the expense of Ni, as shown in Figure 4.13. The maximum Ni_3Al content obtained in the temperature range of 900 to 1150°C was 52 mol% at 1100°C with the rest being elemental Ni, as shown in Figure 4.21. The maximum calculated Ni_3Al content was 100 mol% at 1100°C, as seen in Figure 4.15(a). Conversion of metal chlorides: The conversion degrees of metal chlorides were determined as functions of temperature by analyzing the total amount of Ni and Al present in the solid product. The entire amount of the product must be collected in order to obtain a correct degree of conversion. That was difficult experimentally, because some of the solid product was lost during washing, filtering, and collecting the dried product particle from the filter paper. The amount of powder product after each run itself was small, in the range of 1 to 3 g. Thus, the calculated conversion values of chlorides have about 10 to 15% error. The variation of conversion degrees of chlorides with temperature is presented in Table 4-6. The experimental conditions were the same as those presented in Figure 4.21. A maximum conversion value of 94% of $NiCl_2$ was achieved. Particle size and shape: It was difficult to isolate the fine particles because they were quite agglomerated. Suspending the powder in water in an ultrasonic bath for an hour resulted in separated particles in some cases. The phases of Ni, Ni_3Al, $NiAl_3$ and some NiAl were identified from the electron diffraction pattern. The size and shape of individual particles in the agglomerated samples were difficult to determine.

To summarize, the hydrogen reduction of $AlCl_3$ and $NiCl_2$ mixtures to form nickel aluminides required the presence of aluminum vapor to initiate the reduction reaction. The fact that aluminum chloride is reduced by this reaction scheme is very significant, because the reduction of $AlCl_3$ alone by H_2 is thermodynamically unfavorable at the moderate temperatures of this work. Aluminum vapor, even when present at a very small partial pressure, promoted the reduction of $AlCl_3$ by forming fine nuclei of the aluminide which is difficult to form on the particles of nickel alone.

The amount of an intermetallic phase in the product could be maximized by controlling the reactant partial pressures and temperature. The maximum percentage of the reduced Ni contained in Ni_3Al was 80 % at 1050°C and the H_2, $AlCl_3$, and $NiCl_2$ partial pressures of 57, 1.5, and 0.5 kPa, respectively. The conversion of the limiting chloride $NiCl_2$ was as high as 95%. The particles of Ni_3Al obtained were in the size ranges of 50 to 100 nm. The X-ray diffraction pattern of solid product showed some peaks that matched some of the peaks of NiAl and $NiAl_3$ at the high $AlCl_3/NiCl_2$ partial pressure ratios of 12 and 18. Very fine and extensively agglomerated product particles were observed by TEM. From the electron diffraction pattern, the particles of Ni, Ni_3Al, $NiAl_3$, and NiAl were identified. Their size was difficult to determine.

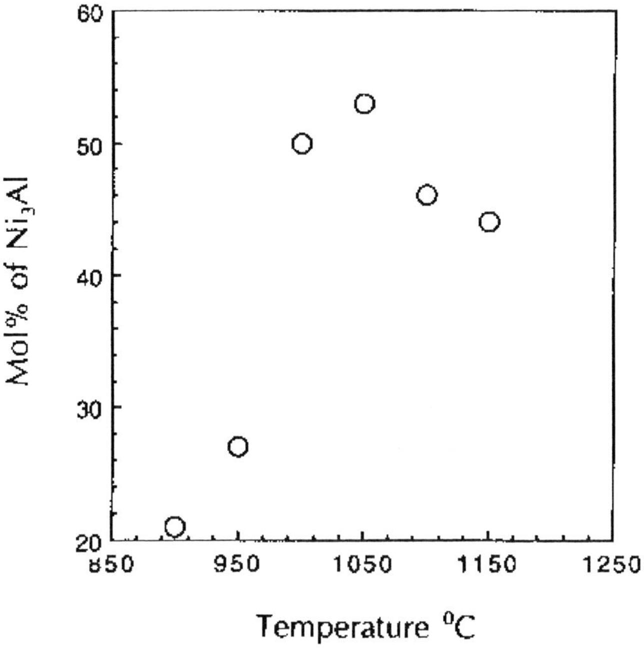

Figure 4.21. Ni_3Al content in the solid product versus temperature. (Experimental conditions are partial pressures of H_2, $AlCl_3$, and $NiCl_2$ = 57, 1.4, and 0.5 kPa, respectively; total flow rate = 4.5 L/min at 25°C and 86.1 kPa.) [Adapted from Sohn and PalDey (1998d)].

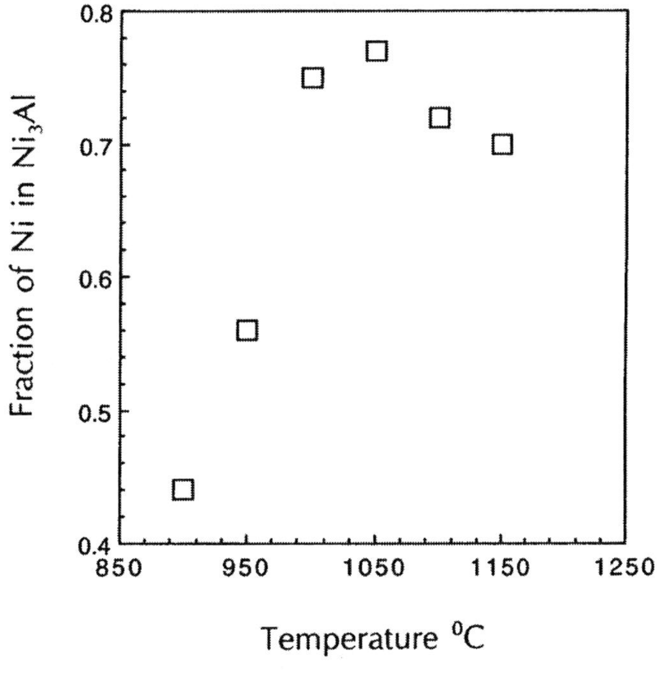

Figure 4.22. The fraction of reduced Ni in Ni_3Al versus temperature. [Adapted from Sohn and PalDey (1998d)].

Table 4.6. Conversion of $NiCl_2$ and $AlCl_3$. (Partial pressures of $AlCl_3$, $NiCl_2$, and H_2 = 1.4, 0.5, and 57 kPa, respectively)

Temperature (°C)	Conversion of $NiCl_2$ (%)	Conversion of $AlCl_3$ (%)
900	57	2.5
950	63	3.7
1000	94	7.8
1050	82	7.7
1100	64	5.1
1150	68	7.3

4.4. Other Intermetallic and Alloy Powders

Synthesis of Fe-Co Powders

Sohn et al. [2004] applied the chemical vapor synthesis (CVS) process to the preparation of Fe-Co powder. In this process, a vapor-phase mixture of $FeCl_3$ and $CoCl_2$ was reduced by hydrogen according to the following reaction:

$$FeCl_3(g) + xCoCl_2(g) + (1.5 + x)H_2(g) = (3 + 2x)HCl(g) + Fe\text{-}xCo(s) \quad (4\text{-}15)$$

An experimental apparatus similar to that given in Section 3.5 was used in this work. Inside the tubular reactor, a ceramic basket (with small holes at the bottom) filled with small hollow quartz cylinders was positioned at a given temperature zone for rapidly vaporizing the feed metal chloride powder mixture. This assured the composition of the chloride vapor mixture to be the same as that of the fed powder. At the bottom of the reactor, a layer of filter cloth was used to collect powder products.

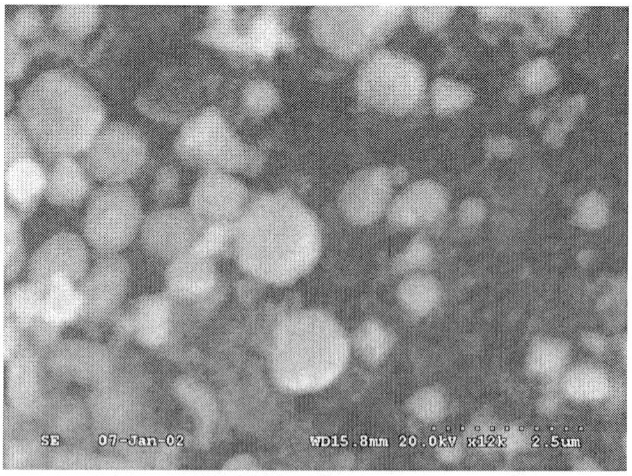

Figure 4.23. SEM micrograph of product powder. [Adapted from Sohn et al. (2004)].

Anhydrous ferric chloride, $FeCl_3$, (>98%, from EM Science) and cobalt chloride, $CoCl_2$, (97%, from Aldrich Chemical Company, Inc.) were used as the reactants. The metal chlorides were ground, weighed, and mixed at a desired molar ratio in a glove box under an argon atmosphere. The mixture was charged into the powder feeder in a gas-tight box.

The measured feeding rate of the metal chloride mixture of 50/50 mole percent $FeCl_3$ and $CoCl_2$ was 1.1 g/min. The flow rate of hydrogen gas was 16.5 NmL/s. Thus, the H/Cl atomic ratio was 4, the hydrogen flow rate being 4 times the stoichiometric requirement. The flow rate of argon was the same as that of hydrogen. The furnace temperature controller was set to 1100°C and the total pressure was 86.5 kPa, the atmospheric pressure at Salt Lake City.

The produced powder was first washed with distilled water to remove the unreacted metal chlorides, if any, washed with acetone to remove water, and then dried under argon. The washed samples were analyzed by a scanning electron microscope (SEM) (Model: HITACHI S-3000N) equipped with EDS (energy dispersive spectrometer).

Figure 4.23 shows an SEM micrograph of product powder. The apparent particle size ranged from 200 to 1000 nm, but each particle, especially the larger ones, were an aggregate of smaller crystals. The Co/Fe ratio in the washed powder ranged from 3.3 to 3.7, although the ratio in the feed chloride mixture was 1. This and other evidence such as the high Fe contents in the unwashed samples that contains the unreacted chlorides indicate that $CoCl_2$ is more readily reduced by hydrogen than $FeCl_3$. The temperature at which the samples were produced was in the range 900 - 1000°C, which would place the product in the α–phase or γ–phase region, depending on the composition [Nishizawa and Ishida, 1993].

Synthesis of Ni_4Mo Powders

Ni_4Mo, an ordered alloy, is of interest because of its potential as a high temperature structural material [Westbrook, 1967]. Sohn and PalDey [1998c] synthesized ultrafine particles of Ni_4Mo by reducing mixtures of $NiCl_2$ and $MoCl_5$ vapors by hydrogen. The effects of $MoCl_5/NiCl_2$ and $H_2/NiCl_2$ partial pressure ratios and temperature on the product composition were investigated. The maximum Ni_4Mo content obtained was 93 mol%. Product particles were in the size range of 100–200 nm. A coating of Ni_4Mo + Ni of 0.7 μm thickness on a Ni substrate was also prepared.

The apparatus and experimental procedure described in Section 4.1 were used in this work. The product particles were washed at room temperature with dilute HCl solution followed by water to remove unreacted chlorides. The samples were dried under a flowing argon atmosphere. Ni4Mo was formed along with elemental Ni and/or Mo. The chemical reactions involved are:

$$4NiCl_2(g) + MoCl_5(g) + 6.5H_2 = Ni_4Mo(s) + 13HCl(g) \quad (4\text{-}16)$$

$$NiCl_2(g) + H_2 = Ni(s) + 2HCl(g) \quad (4\text{-}17)$$

$$MoCl_5(g) + 2.5H_2 = Mo(s) + 5HCl(g) \quad (4\text{-}18)$$

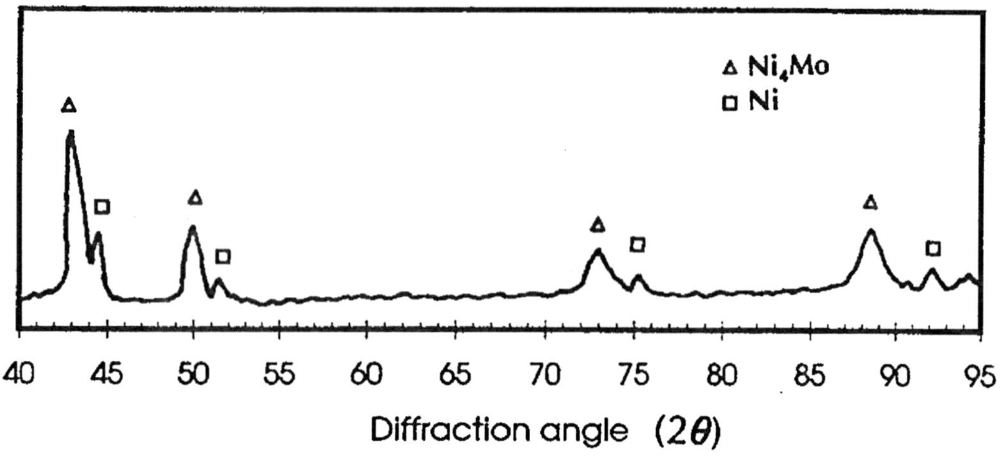

Figure 4.24. Sample X-ray diffraction pattern of the solid product (input p_{MoCl5} = 0.32 kPa; input p_{NiCl2} =1.53 kPa; input p_{H2} = 25 kPa; T= 1000°C). [Adapted from Sohn and PalDey (1998c)].

The Ni_4Mo formation reaction proceeded without any nucleation promoters, because the hydrogen reduction of nickel chloride and molybdenum chloride is favorable individually.

Preliminary experiments were carried out and the solid product particles were analyzed by XRD, an example of which is shown in Figure 4.24. The solid products were relatively well crystallized. In this work the metal chlorides were vaporized at different temperatures [PalDey, 1995]. Results of these experiments and the observation from the XRD of the solid product are presented in Table 4-7. In all these experiments, the flow rates of carrier argon, purge argon, and H_2 were kept, respectively, at 1.0, 0.5, and 2.0 L/min at 25°C and 86.1 kPa. The reaction temperature was 1005°C and the $NiCl_2/MoCl_5$ vapor pressure ratio was varied from 1.9 to 28. At the lower range of the ratio, product phase contained elemental Mo and Ni_4Mo. The peak intensity of Mo decreased with increasing partial pressure ratio up to 4.8 where the product contained elemental Ni, Ni_4Mo and Mo. Beyond the ratio of 9.6, the product phase contained Ni and Ni_4Mo with no Mo. The peak intensity of Ni_4Mo increased at the expense of Ni at a higher ratio of 27.9. At this point, the vapor pressures of both chlorides were increased in order to increase the concentration of the chlorides in the gas phase and to enhance the nucleation and growth of the product particles. The last two rows of results in the table present the result of increased vapor pressures of both chlorides. A significant increase in the peak intensity of the intermetallic compound was obtained.

Quantitative measurements of the solid product were then carried out while varying the reactant partial pressures and temperatures. The Ni_4Mo content in the product was computed from the elemental analysis for those cases in which the XRD result showed the presence of only two solid phases. Thus, if the XRD patterns showed Ni_4Mo and Mo phases, the Ni_4Mo content was calculated from that of nickel with the surplus Mo giving the content of the elemental molybdenum. This gives a much more quantitative composition than from XRD patterns, although this method cannot be applied to samples containing three phases. The composition of such a product was expressed by the following two quantities:

1) The mol% of Ni_4Mo in the product, and
2) Fraction of reduced Ni in Ni_4Mo.

Table 4.7. Results of the qualitative experiments with the Ni–Mo system

Vaporizer temperature		Equilibrium vapor pressure of $MoCl_5$ (kPa)	Ratio of vapor pressures $NiCl_2/MoCl_5$	XRD observation
$MoCl_5$ (°C)	$NiCl_2$ (°C)			
170	825	2.47	1.9	Strong peaks of Mo, weak peaks of Ni_4Mo
160	825	1.32	3.56	Strong peaks of Mo, weak peaks of Ni_4Mo
150	800	0.69	4.81	Peaks of Mo, Ni_4Mo and Ni
145	800	0.49	6.76	Peaks of Mo, Ni_4Mo and Ni
140	800	0.35	9.56	Strong peaks of Ni and Ni_4Mo
130	800	0.17	19.7	Strong peaks of Ni_4Mo, weak peaks of Ni
130	825	0.17	27.9	Strong peaks of Ni_4Mo, weak peaks of Ni
160	900	1.32	19.9	Very strong peaks of Ni_4Mo, weak Ni, Mo
155	900	0.95	27.7	Very strong peaks of Ni_4Mo, weak Ni

Effect of reactant partial pressures and temperature: Figure 4.25 shows the variation of the Ni_4Mo content with the $MoCl_5/NiCl_2$ partial pressure ratio at constant H_2 and $NiCl_2$ partial pressures of 49 and 1.5 kPa, respectively. The temperature was 1000°C and the total flow rate was kept at 3.5 L/min at 25°C and 86.1 kPa.

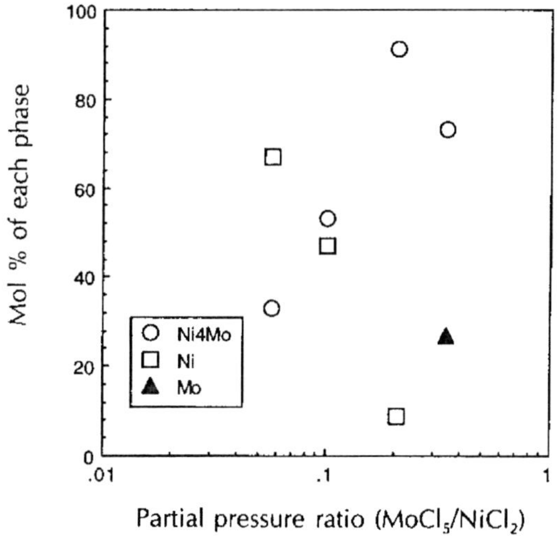

Figure 4.25. Composition of the solid product versus the input $MoCl_5/NiCl_2$ partial pressure ratio (input p_{H2} = 49 kPa; input p_{NiCl2} =1.5 kPa; T =1000°C; total gas flow rate = 3.5 L/min at 25°C and 86.1 kPa). [Adapted from Sohn and PalDey (1998c)].

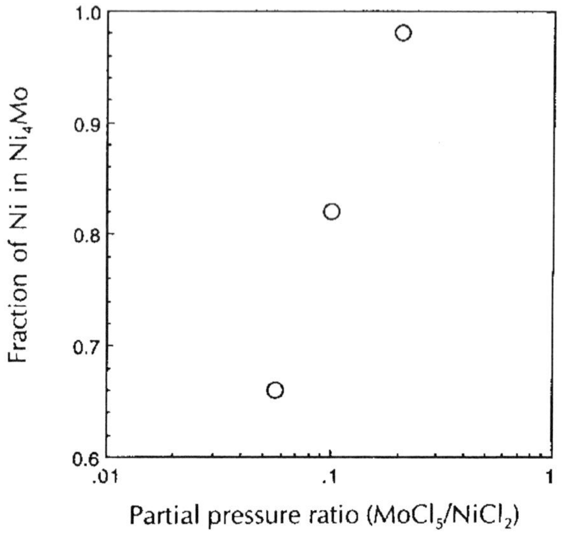

Figure 4.26. The fraction of reduced Ni in Ni_4Mo versus the input $MoCl_5/NiCl_2$ partial pressure ratio. (The same experimental conditions as for Figure 4.25) [Adapted from Sohn and PalDey (1998c)].

The Ni_4Mo content increased from 33 to 91 mol% as the $MoCl_5/NiCl_2$ ratio increased from 0.06 to 0.2. The rest of the product was elemental Ni. The product contained Ni_4Mo along with elemental Ni and Mo beyond the ratio of 0.2. At the ratio of 0.35, the product composition was 73 mol% Ni_4Mo and 27 mol% elemental Mo. These quantitative results agreed well with the XRD results presented in Table 4-7. The fraction of reduced Ni contained in the intermetallic phase versus the input ratio of $MoCl_5/NiCl_2$ is shown in Figure 4.26. Starting from 0.66, the fraction of Ni in Ni_4Mo increased to 0.98. The rest of Ni was present as elemental Ni.

The variation of the Ni_4Mo content with the input $H_2/NiCl_2$ partial pressure ratio was determined at 1000°C and at constant $MoCl_5$ and $NiCl_2$ partial pressures of 0.32 and 1.5 kPa, respectively. The product contained Ni_4Mo and Ni phases. The Ni_4Mo content increased to 92 mol% as the $H_2/NiCl_2$ ratio was increased from 8 to 32, as can be seen in Figure 4.27. Figure 4.28 shows the fraction of reduced Ni contained in Ni_4Mo, which increased from 0.91 to 0.98 in the same range of the $H_2/NiCl_2$ ratio.

The effect of temperature on the composition of solid product was studied in the range of 900–1100°C. The total flow rate was 3.5 L/min at 25°C and 86.1 kPa. The input $MoCl_5$, $NiCl_2$ and H_2 partial pressures were 0.32, 1.53 and 25 kPa, respectively. Starting from 80 mol% at 900°C, the Ni_4Mo content reached a maximum of 93 mol% at 1100°C as shown in Figure 4.29. The rest of the product was elemental Ni. The variation of the fraction of Ni contained in Ni_4Mo with temperature is shown in Figure 4.30, the maximum value being 0.98 at 1100°C.

Particle size and shape: Figure 4.31 presents the TEM micrographs of the powder product. Figure 4.31(a) shows a chain of particles of the size range 100–200 nm. At a higher magnification, some hexagonal and rectangular particles were observed. Figure 4.31 (b) shows rectangular particles of 100 × 125 nm size, and Figure 4.31(c) shows hexagonal particles of 130 × 170 nm size [Sohn and PalDey, 1998c]. In principle, the lattice parameters from the diffraction patterns of the separated particles should make it possible to identify each particle as elemental Ni or intermetallic Ni_4Mo. Ni and Ni_4Mo, however, have the same FCC unit cells and they have almost the same lattice parameters of 0.352 and 0.355 nm, respectively. From the diffraction pattern, the measured lattice parameter was 0.355 nm.

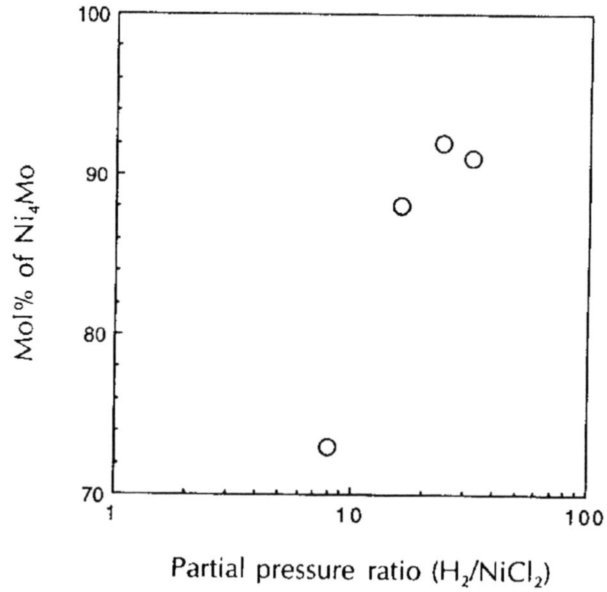

Figure 4.27. Ni_4Mo content in the solid product versus the input $H_2/NiCl_2$ partial pressure ratio (input p_{MoCl5} = 0.32 kPa; input p_{NiCl2} = 1.5 kPa; T= 1000°C; total gas flow rate = 3.5 L/min at 25°C and 86.1 kPa). [Adapted from Sohn and PalDey (1998c)].

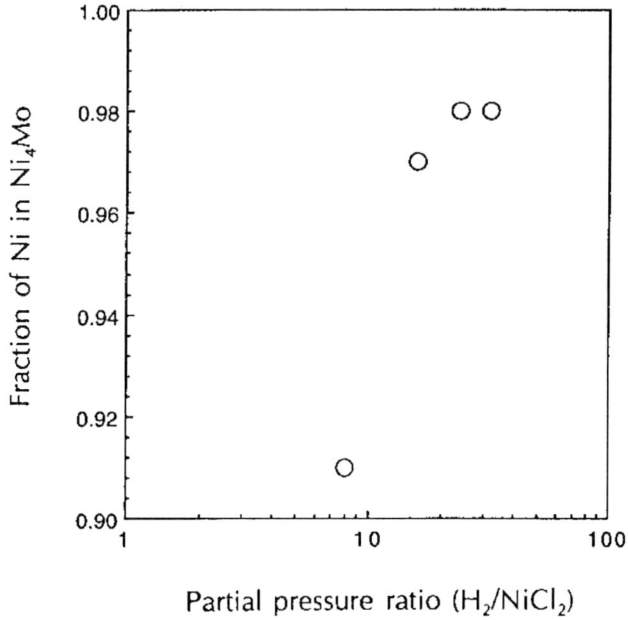

Figure 4.28. The fraction of reduced Ni in Ni_4Mo versus the input $H_2/NiCl_2$ partial pressure ratio. (The same experimental conditions as for Figure 4.27) [Adapted from Sohn and PalDey (1998c)].

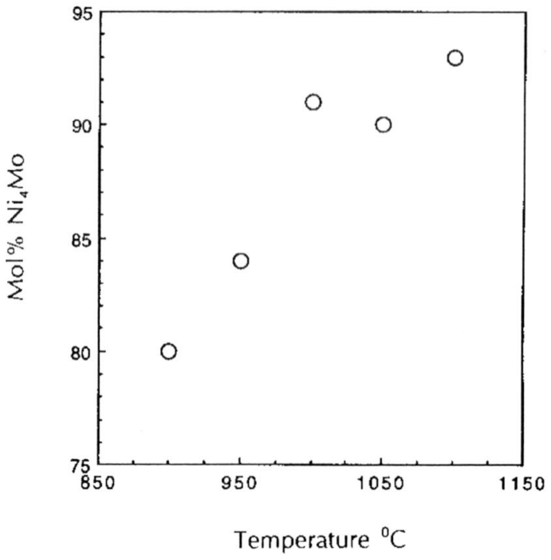

Figure 4.29. Ni_4Mo content in the solid product versus temperature (input p_{NiCl2} =1.53; input p_{MoCl5} = 0.32; input p_{H2} = 25 kPa; total gas flow rate = 3.5 L/min at 25°C and 86.1 kPa). [Adapted from Sohn and PalDey (1998c)].

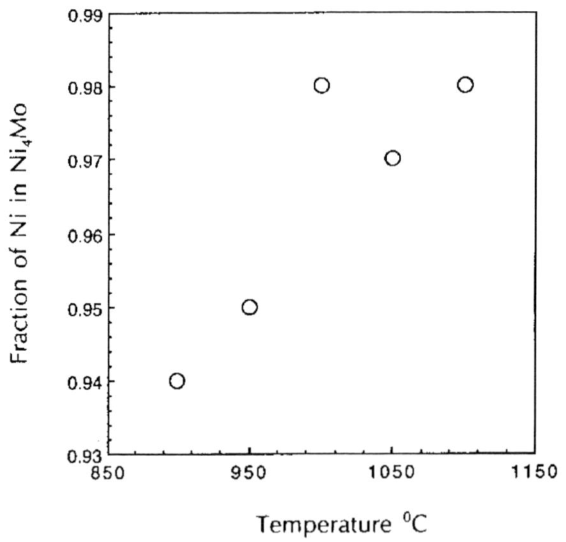

Figure 4.30. The fraction of reduced Ni in Ni_4Mo versus temperature. (The same experimental conditions as for Figure 4.29) [Adapted from Sohn and PalDey (1998c)].

Preparation of Ni_4Mo Coating on Nickel Substrate

A qualitative investigation on the preparation of a coating of Ni_4Mo on a metal substrate was carried out. Small pieces of Ni wire substrates of diameter 2.0 mm were placed on a substrate holder which was basically a quartz boat with grooves to hold the substrates. The $NiCl_2$, $MoCl_5$ and H_2 partial pressures in these experiments were 1.53, 0.32

and 25 kPa, respectively. The reaction temperature was 1005°C and the total flow rate was constant at 3.5 L/min at 25°C and 86.1 kPa. The wires were coated for 30 min. After the experiment, the reactor was cooled to room temperature with a constant flow of argon. The coated wires were then subjected to XRD and electron microscopy. The XRD of the coated surface showed the presence of Ni_4Mo. The substrates were cut, mounted in epoxy resin, polished and observed by SEM, as shown in Figure 4.32. Microprobe analysis showed that both Ni_4Mo and Ni were present in the coated region. The thickness of the coated region was about 0.7 μm [Sohn and PalDey, 1998c].

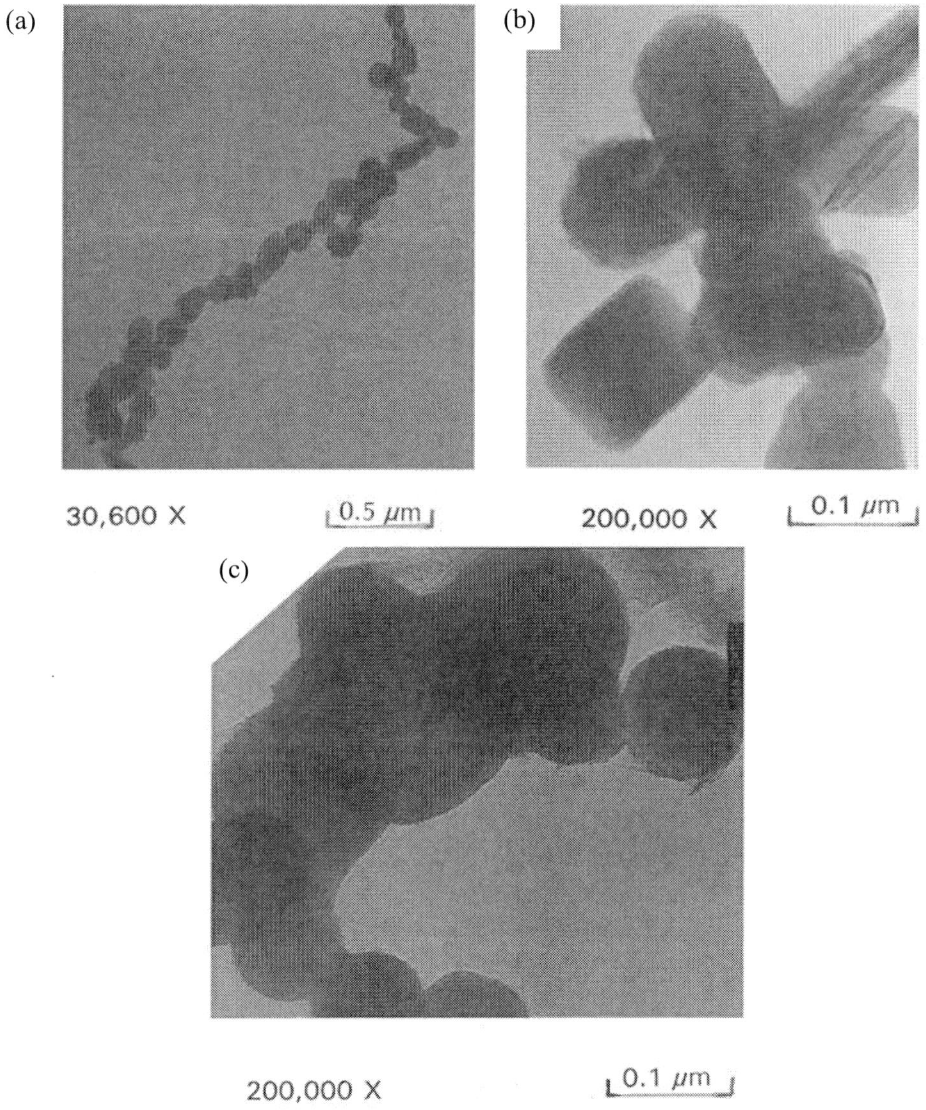

Figure 4.31. TEM micrographs of the solid product particles. (a) a chain of particles; (b) rectangular particles; (c) hexagonal particles. [Adapted from Sohn and PalDey (1998c)].

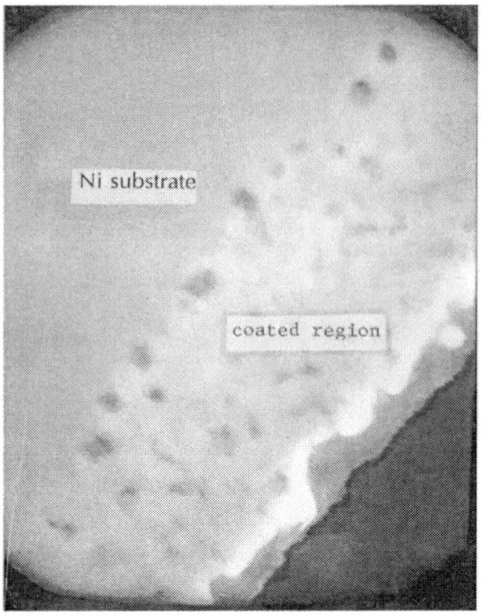

Figure 4.32. SEM micrograph of the Ni_4Mo coating on Ni substrate. [Adapted from Sohn and PalDey (1998c)].

In summary, the co-reduction of $NiCl_2$ and $MoCl_5$ vapors by hydrogen produced Ni_4Mo along with some Ni and/or Mo. No other intermetallic phases were observed in the experimental conditions tested. The effects of $MoCl_5/NiCl_2$ and $H_2/NiCl_2$ partial pressure ratios and temperature on the Ni_4Mo content in the product solid were studied. By carefully controlling the chloride partial pressures and temperature, the Ni_4Mo phase with only a small amount of an elemental metal was obtained. The maximum Ni_4Mo content obtained was 93 mol% at 1100°C and $NiCl_2$, $MoCl_5$, and H_2 partial pressures of 1.53, 0.32, and 25 kPa, respectively. The rest of the product was elemental Ni. The maximum fraction of reduced Ni contained in Ni_4Mo was 0.98. Fine product particles of Ni and Ni_4Mo observed with TEM had rectangular or hexagonal shapes and were in the size range of 100–200 nm. A qualitative study on the preparation of Ni_4Mo coating on a Ni substrate was also carried out. A coated layer of thickness 0.7 μm containing Ni_4Mo was formed on the Ni substrate.

Chapter 5

TUNGSTEN CARBIDE NANOPOWDER

ABSTRACT

The preparation of tungsten carbide nanopowder by the chemical vapor synthesis (CVS) process using volatile and non-volatile precursors is described in this chapter. The use of a non-volatile precursor requires a plasma application, which provides an extremely high temperature to volatilize normally non-volatile materials such as stable oxides.

Tungsten carbide has many industrial applications, thanks to its much desired properties of high hardness and good wear resistance. It is used in tools for the metalworking and the drilling and mining industries under high pressure, high temperature, and corrosive environments. Mechanical properties such as hardness, compressive strength, and transverse rupture depend on the composition and microstructural parameters such as grain size [Lee and Kang, 2006; Fang et al., 2005; Petersson and Ågren, 2004; Upadhaya, 2002]. Previous investigations [Zawrah, 2007; Shi et al., 2006; Nersisyan et al., 2005; Wu et al., 2004; Fu et al., 2001; Wahlberg et al., 1997; Zhu and Manthiram, 1996] have shown that the reduction of tungsten carbide grain size results in a significant improvement of the mechanical properties. Thus, nanosized tungsten carbide powder has a considerable potential.

For the preparation of fine tungsten carbide powder, compounds such as tungsten hexachloride (WCl_6) [Hojo et al., 1978], tungsten hexafluoride (WF_6) [Fitzsimmons and Sarin, 1995], and tungsten hexacarbonyl [$W(CO)_6$] [Kim and Kim, 2004] are generally used as precursors based on their relatively low evaporation temperatures as well as their high reducibility by hydrogen. As reducing and carburizing agents, hydrocarbons such as methane (CH_4) [Medeiros et al., 2001; Leclercq et al., 1996; Kelly et al., 1992], propane (C_3H_8) [Tang et al., 1995], and acetylene (C_2H_2) [Won et al., 1993] have been used. Methane is the most commonly used carburizing agent because it is most readily available and stable up to a high temperature [Gao and Kear, 1995].

In recent years, the development of nanosized tungsten carbide powder has attracted much attention based on the expectation of significant improvements in the mechanical properties, such as hardness, compressive strength and rupture strength, of cemented tungsten carbide that retains nanograin structure. The reduced scale of the structure also means an increase in the dislocation density and the amount of grain boundaries per unit volume.

Furthermore, nanosized powders start sintering at lower temperatures and most of densification occurs while they are in solid state [Malewar et al., 2007; Fang et al., 2005; Wahlberg et al., 1997; Zhu and Manthiram, 1996; Rabe and Wäsche, 1995].

The preparation of nanosized tungsten carbide powder by chemical vapor synthesis (CVS) will be discussed below.

5.1. SYNTHESIS FROM TUNGSTEN CHLORIDE

Early Investigations

Hojo et al. [1978] synthesized nanosized tungsten carbide powder from the WCl_6-CH_4-H_2 reaction system. The reactor was an alumina tube of 2.2 cm inner diameter placed in a horizontal furnace consisting of sublimation furnace and reaction furnace. The effective length of the reaction furnace was 20 cm. Tungsten hexachloride powder was placed in the sublimation furnace kept at 400°C and nitrogen was used as the carrier gas. Hydrogen and methane were used as the reducing and carburizing agent, respectively. Two methods of the introducing the WCl_6 vapor into the reaction zone were tested. Method 1 involved mixing all gas streams in the sublimation zone at 400°C before entering the reactor. Method 2 involved mixing the gas streams in the high temperature zone at a temperature from 1000°C to 1400°C. Experiments were carried out to determine the effects of reaction temperature on the product composition and particle size. The effect of methane concentration on the phase composition was also investigated. The results indicated that the degree of carburization increased as the reaction temperature and methane concentration increased. The product was mostly WC in method 1 at the reaction temperature of 1400°C. However, more W_2C was formed in method 2 than in method 1 under otherwise the same conditions. Particle size was affected by the temperature where the gases were mixed, with 40~50 nm from method 1 and 80~110 nm from method 2. Substantial WC formation started to occur above 1000~1200°C.

Tang et al. [1995] synthesized nanosized tungsten carbide particles from WCl_6-C_3H_8-H_2 mixtures. The reactor was an alumina tube of 2.8 cm inner diameter with an effective length of 15 cm placed in a vertical furnace. Tungsten hexachloride powder was placed in a sublimation tube maintained at 150 to 200°C, and a hydrogen-propane mixture was passed through the tube into the reaction zone where temperature was varied from 1280 to 1500°C. The product was WC with some W_2C and W. A post-treatment of synthesized powder by hydrogen fully removed the 9.2 wt % excess carbon in the product after 1 hour of treatment at 1000°C.

Won et al. [1993] synthesized nanosized tungsten carbide powder from WCl_6-C_2H_2-H_2 mixtures. The reactor was an alumina tube of 3 cm inner diameter and 60 cm length placed in a horizontal furnace. The precursor was introduced into the reactor from a vaporizer. Hydrogen and acetylene were used as the reducing and carburizing agent, respectively. These reactant gases mixed with argon flowed through the vaporizer to carry the WCl_6 vapor. The experimental conditions were as follows: Reaction temperature was varied from 1000°C to 1200°C, the vaporizer temperature was varied from 250°C to 350°C, and H_2 flow rate was varied from 1 L/min to 4 L/min with the C_2H_2 flow rate of 0.2 L/min. Under most experimental conditions, a mixture of α-WC and β-W_2C was produced. The particle size increased as the reaction temperature and WCl_6 vaporizer temperature increased but it decreased with increasing H_2 flow rate. Particle size varied from 100 to 1000 nm depending on experimental parameters. [It is noted that a particle can be made up of polycrystalline grains or just a single crystalline grain.]

It is seen from the scanning electron microscope (SEM) photomicrographs shown in Figure 5.1 that the particles produced at 1273 K were spheroidal with relatively uniform diameters of 100-200 nm. At 1373 K, larger particles of 200-300 nm diameters were produced. At a still higher temperature of 1473, however, the diameters of the product particles decreased to below 100 nm. At all temperatures, the product particles were somewhat agglomerated.

The particles produced at 1273 K were also examined by transmission electron microscopy (TEM). Figure 5.2(a) shows spheroidal particles that form chains. Particles produced at other temperatures showed similar chain formation. The product, however, also contains rod-shaped particles of about 100 nm width and 400-500 nm length, as shown in Figs. 5-2(b) and 5-2(c). The electron diffraction pattern of this product is shown in Figure 5.2(d), indicating the crystallinity of the material.

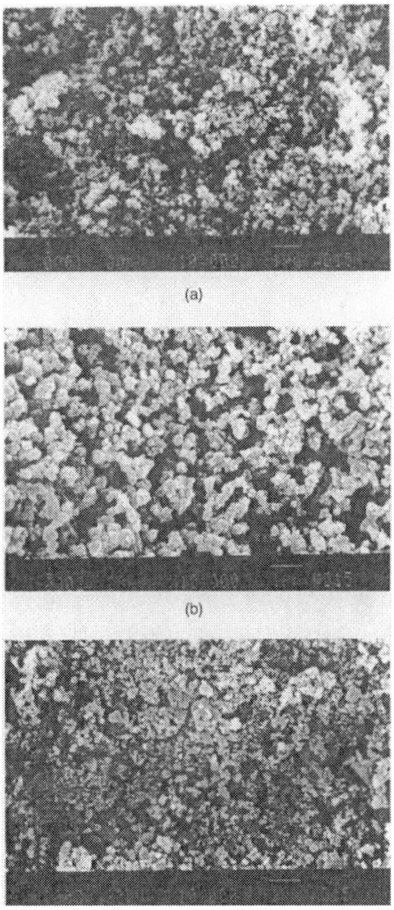

Figure 5.1. SEM photomicrographs of the products obtained at various reaction temperatures (H_2 flow rate = 2 L/min; C_2H_2 flow rate = 0.2 L/min; WCl_6 vaporizer temperature = 573 K). (a) 1000°C (1273 K), (b) 1100°C (1373 K), and (c) 1200°C (1473 K). [Adapted from Won et al. (1993)].

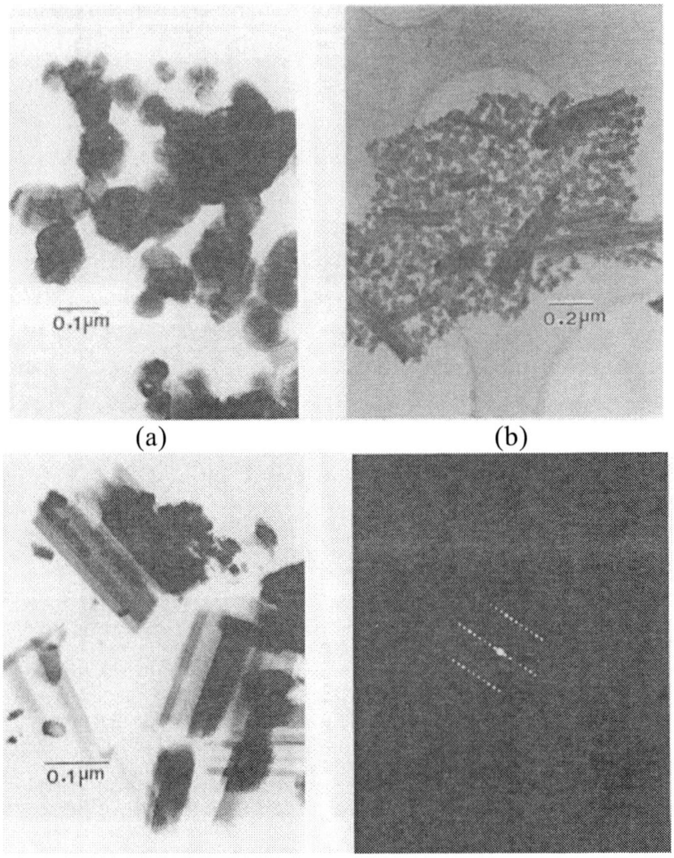

(a) (b)

(c) (d)

Figure 5.2. (a – c) TEM photomicrographs and (d) diffraction pattern of the product (reaction temperature = 1273 K). [Adapted from Won et al. (1993)].

The grain sizes of the particles were calculated from X-ray diffraction data on the (100) face by applying the Scherrer equation described in Section 3.2. These results, together with the nitrogen BET surface areas, are given in Figure 5.3. It is seen that the particle size of the product increases with reaction temperature up to 1373 K and then decreases with temperature. The change in the surface area mirrors that in the particle size. This behavior is consistent with the results of the electron microscopy analysis shown in Figure 5.1.

The effect of the vaporizer temperature was tested by varying it from 523 to 623 K. At the vaporizer temperature of 523 K, the particles were spheroidal and non-uniform in size in the range 200-1000 nm. When the vaporizer temperature was raised to 573 K, the particle size decreased to relatively uniform values of 200-300 nm. Particle growth in the reactor was faster at the vaporizer temperature of 623 K, resulting in an increased particle size in the range of 200-600 nm. Their TEM photomicrographs were essentially the same as that in Figure 5.2(a).

The effect of H_2 flow rate was tested by varying it from 1 to 4 L/min. At the lowest flow rate, the product contained a mixture of spheroidal particles, whiskers, needles, and rods. The particles became quite uniform at 2 L/min H_2 flow rate, but again became somewhat non-uniform in size at 4 L/min. The grain size decreased and the surface area increased with H_2 flow rate.

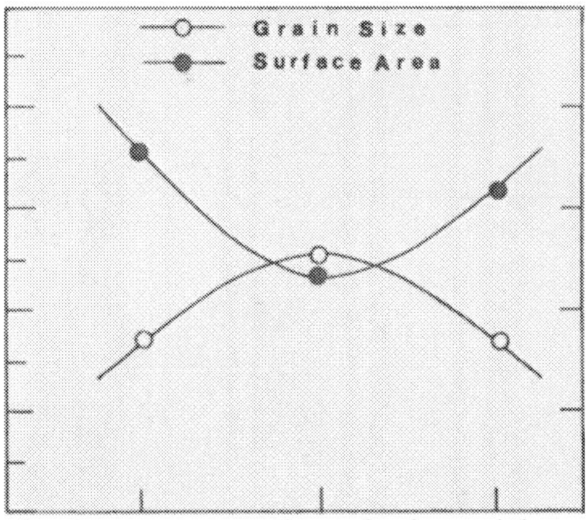

Figure 5.3. Effect of reaction temperature on the grain size and the surface area of the product particles. (Other conditions were the same as in Figure 5.1.) [Adapted from Won et al. (1993)].

Recent Work on Nano-Sized Powder

Sohn and coworkers [Sohn et al., 2007; Ryu et al., 2008c] synthesized tungsten carbide powder from WCl_6-CH_4-H_2 mixtures with small grain sizes. The tubular reactor system used in their work consisted of two powder feeders, a vertical reactor, powder collectors, an off-gas scrubber containing a 5 wt % NaOH solution, and a Bunsen burner, as shown in Figure 5.4. The reactant feeding system and procedure have been described in Section 3.2. A stainless steel tube of 3.2 mm inner diameter was used as the delivery line connected from the powder feeder to the volatilizer. The volatilizer was made of an alumina tube of 6.4 mm inner diameter with holes near the closed end to provide an outlet for the vaporized precursors into the reactor. WCl_6 (99.9%) was used as the precursor, together with hydrogen (99.9%) and methane (99.9%) as the reducing and carburizing agents, respectively. Argon gas (99.9%) was passed through the powder feeding system as the carrier gas as well as to flush the reactor. Methane was introduced through the delivery line into the WCl_6 volatilizer to facilitate mixing with volatilized WCl_6. The WCl_6 volatilizer temperature was sufficiently low to prevent the reduction and/or carburization of WCl_6 by methane in the volatilizer, which would result in a deposition of solids in the volatilizer. Hojo et al. [1978] reported that substantial WC formation from WCl_6 by methane starts above 1000~1200°C, as also observed in the work discussed here. The reactor was an alumina tube of 54 mm inner diameter and 1.5 m length and placed in a vertical furnace. The produced powder was collected using a Teflon-coated polyester filter with a 1 μm pore size. A mixture of argon, hydrogen, and methane gases was fed into the reactor to reduce the reactant precursors and carburize the reduced tungsten. The reactor was purged with an Ar flow of 2 L/min (25°C, 86.1 kPa total pressure at Salt Lake City) for 20 minutes before and after each experiment.

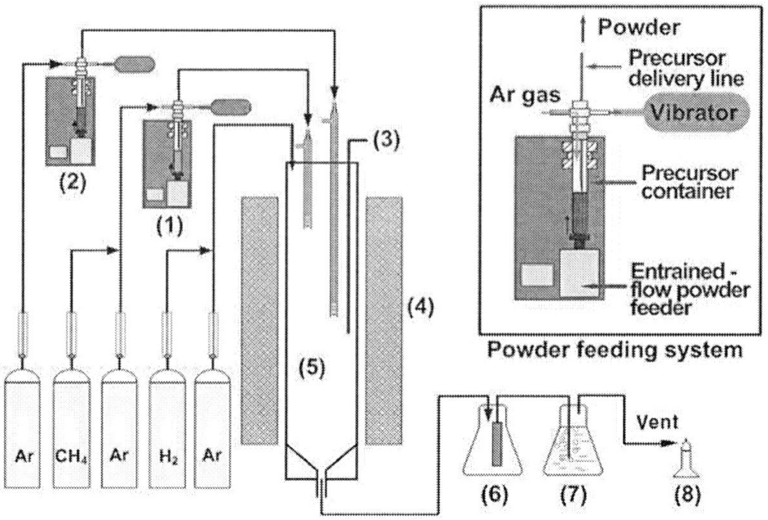

Figure 5.4. Schematic diagram of the tubular reactor system: (1) entrained-flow powder feeder for WCl_6, (2) standby entrained-flow powder feeder, (3) thermocouple, (4) vertical furnace, (5) alumina reactor, (6) powder collector, (7) scrubber, and (8) Bunsen burner. [Adapted from Ryu et al., 2008d].

According to Hojo et al. [1978], WCl_6 is first reduced to metallic tungsten which in turn reacts with methane to form W_2C followed by carburization to WC. It is thus expected that the extent of carburization will increase with increasing methane concentration and temperature. Thus, the apparatus was designed to promote the production of WC over that of W_2C by introducing CH_4 together with WCl_6 through the precursor delivery line. This facilitated the mixing of the two reactants, and the amount of W_2C in the product was minimized, as shown in Figure 5.5(a). Figure 5.5 also shows the XRD patterns of the products obtained at different reaction temperatures. The main product was WC but some W_2C formed with a decrease in reaction temperature, in agreement with a thermodynamic analysis presented by Hojo et al. [1978].

The grain size of WC was calculated from the XRD pattern by applying the Scherrer equation calibrated by using standard samples [Cullity, 1978], as described in Section 3.2. It is noted that because the powders in this study were produced in the vapor phase, no lattice strain effects were considered in this grain size analysis. As the reaction temperature increased from 1200°C to 1400°C, the degree of carburization increased and the grain size of WC increased from 12 ± 1 to 18 ± 1 nm. Because the amount of W_2C was least at 1400°C, the subsequent experiments were performed at this temperature.

The effect of the CH_4/H_2 ratio on the product composition is shown in Figure 5.6. The amount of W_2C decreased with increasing CH_4/H_2 ratio. The grain size of WC was 18 ± 1 nm at $CH_4/H_2 = 0.2$ and decreased to 12 ± 1 nm when no H_2 was used. CH_4 alone was enough to reduce and carburize WCl_6 to WC.

Tungsten carbide powder is mainly used to make bulk components by consolidation by sintering. Therefore, the presence of the W_2C phase can be tolerated because it can be fully carburized during the sintering process. It can also be fully carburized by a separate post-treatment.

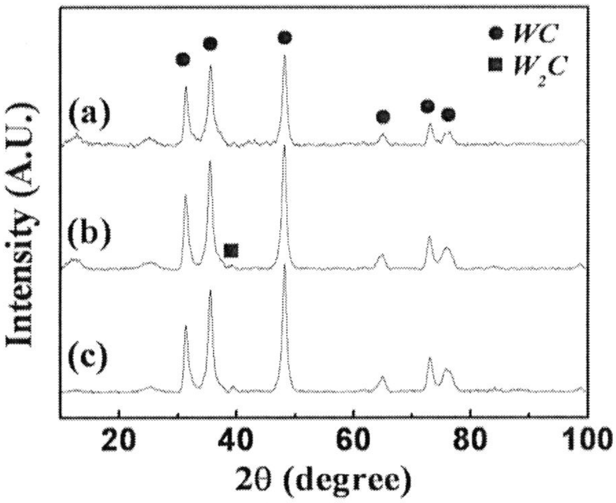

Figure 5.5. X-ray diffraction patterns of the product obtained at (a) 1400°C, (b) 1300°C, and (C) 1200°C under the following conditions: CH_4 feeding rate of 0.1 L/min (25°C, 86.1 kPa), H_2 feeding rate of 0.25 L/min (25°C, 86.1 kPa), total flow rate of 1.1 L/min (25°C, 86.1 kPa), and WCl_6 feeding rate of 0.04 g/min. [Adapted from Sohn et al. (2007)].

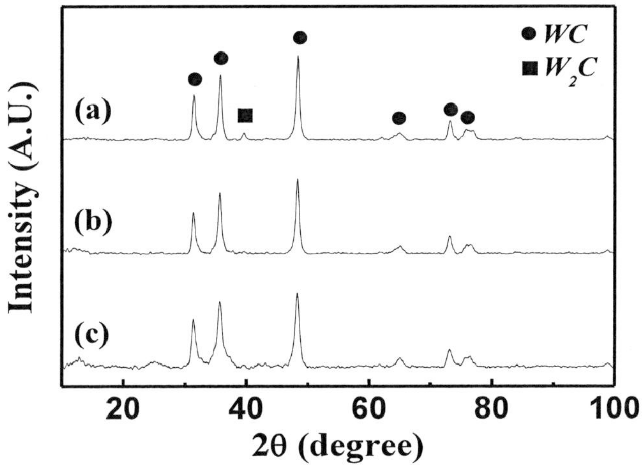

Figure 5.6. Effect of CH_4/H_2 molar ratio. (a) $CH_4/H_2 = 0.2$, (b) $CH_4/H_2 = 0.4$, and (c) only CH_4 under the following conditions: reaction temperature = 1400°C, CH_4 feeding rate = 0.1 L/min (25°C, 86.1 kPa), total flow rate = 1.1 L/min (25°C, 86.1 kPa), and WCl_6 feeding rate = 0.04 g/min. [Adapted from Ryu et al., 2008c].

5.2. Plasma-Assisted Synthesis

From WCl_6-CH_4-H_2 Mixtures

Ryu and co-workers [2008a] synthesized nanosized tungsten carbide powder by a thermal plasma process using WCl_6 as the precursor. The plasma reactor system described in Section

3.2 was used in this work. Tungsten hexachloride of 99.9% purity was used as the precursor. Hydrogen (99.9%) and methane (99.9%) were used as the reducing and carburizing agents, respectively. The produced powder was collected using a Teflon-coated polyester filter with a pore size of 1 μm. The reactor was purged with an Ar flow of 5 L/min (25°C, 86.1 kPa total pressure at Salt Lake City) for 10 minutes before and after each experiment. The product was analyzed by XRD (Siemens D 5000) and the morphology with particle size analysis was done by TEM (JEOL, JEM-3000F). The total carbon content of the powders was measured with a Carbon Determinator (LECO CS-444).

The reduction and carburization of the vaporized precursor by methane-hydrogen mixtures produced nanosized WC_{1-x} powder, which sometimes contained WC and/or W_2C phase. The effects of experimental variables such as the molar ratio of reactant gases, plasma torch power, the flow rate of plasma gas, and the addition of the secondary plasma gas (H_2) were investigated. The product powder had particle sizes less than 20 nm. It was subsequently heated in hydrogen to fully carburize the WC_{1-x}, and W_2C phases to the WC phase as well as to remove excess carbon. Finally, WC powder with particle size less than 100 nm was obtained.

Preliminary experiments: Initial experiments were conducted to find the ranges of conditions at which tungsten carbide could be produced in the WCl_6-H_2-CH_4 system. The experimental conditions in these runs were as follows: the feeding rate of WCl_6 of 3.5 g/min, the flow rate of Ar-H_2-CH_4 mixture to carry the WCl_6 powder of 4 L/min (25°C, 86.1 kPa), the applied plasma torch power of 13 kW, and the flow rate of primary plasma gas (Ar) to generate plasma flame controlled at 38 L/min (25°C, 86.1 kPa) in which no secondary plasma gas (H_2) was used.

The main product under these conditions was WC_{1-x} with small amounts of W_2C and WC phases, as shown in Figure 5.7. This result is consistent with the phase diagram available in the literature [Sara, 1965], which indicates that WC_{1-x} is likely the first stable solid tungsten carbide phase formed as the W-C liquid solution cools towards solidification at 2710°C. After the initial formation the WC_{1-x} phase may be carburized to WC. Further, this phase decomposes into WC and W_2C phases below 2530°C. However, the rapid quenching of the produced particles, which provides little time for either decomposition or further carburization, resulted in the WC_{1-x} phase remaining as the major phase in the product. Figure 5.7 also shows the XRD patterns of the products obtained with different CH_4/H_2 ratios in the feed stream. Crystalline phases were identified by comparing the experimental data with the Joint Committee of Powder Diffraction Standards (JCPDS) files from the International Center for Diffraction Data. The amounts of W_2C and WC decreased with increasing CH_4/H_2 ratio and when CH_4 alone was used as the reaction gas, the product was just WC_{1-x}. The grain size of WC_{1-x} decreased from 13 ± 1 nm at $CH_4/H_2 = 0.5$ to 9 ± 1 nm when no H_2 was used.

These and the subsequent experiments showed that the conversion of WCl_6 to tungsten carbides was complete in the plasma reactor as long as methane was added in excess of the stoichiometric amount. Other factors that significantly affect the WC_{1-x} formation were further tested, as described below.

Effect of methane concentration in the feed stream: The effect of methane concentration in the feed stream was investigated by varying the C/W ratio from 6.3 to 1.5 while maintaining all other conditions constant [WCl_6 feeding rate of 3.5 g/min, flow rate of Ar-CH_4 mixture to carry the WCl_6 powder of 2.5 L/min (25°C, 86.1 kPa) with no H_2, plasma torch power of 13 kW, and flow rate of the primary plasma gas (Ar) of 57 L/min (25°C, 86.1

kPa) in which no secondary plasma gas (H_2) was used]. Figure 5.8 shows the XRD patterns of the products obtained with different C/W ratios in the feed stream. The main product obtained under these conditions was WC_{1-x} with a small amount of W_2C or WC. The formation of WC was suppressed and the presence of W_2C was observed as the methane concentration in the feed stream was decreased. The grain size of WC_{1-x} was 13 ± 1 nm at the C/W ratio of 6.3 and was not affected by decreasing methane concentration in the feed stream within the range tested. The weight percent of carbon in the product was converted to percent (%) excess carbon, which is defined as the percentage of the excess carbon over the corresponding stoichiometric amount if all W was present as WC. The % excess carbon in the product obtained at a C/W ratio of 6.3 in the feed stream was 219 %. It decreased to 26 % as the C/W ratio in the feed stream decreased to 1.5. The presence of free carbon was still observed even at the lowest C/W ratio in the feed stream. However, the free carbon can be removed by post-treatment using hydrogen gas as discussed in the following section.

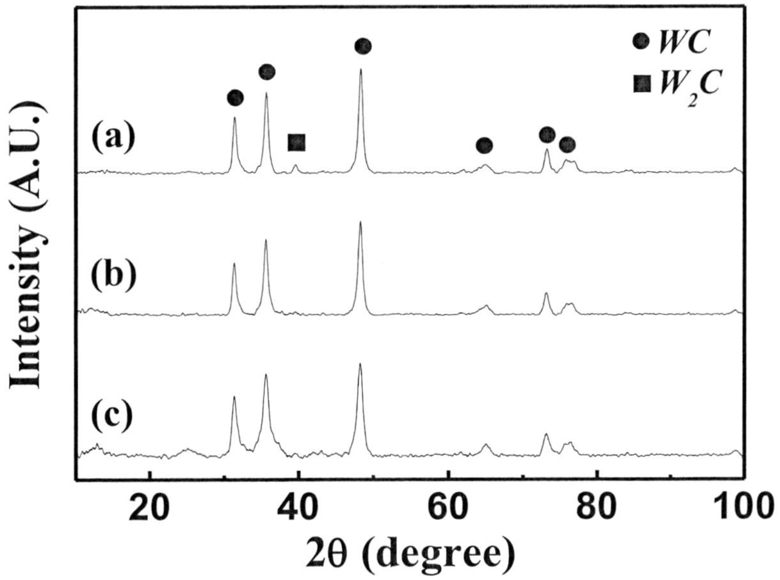

Figure 5.7. X-ray diffraction patterns of the products obtained from different CH_4/H_2 molar ratios at (a) $CH_4/H_2 = 0.5$, (b) $CH_4/H_2 = 1$, and (c) only CH_4 under the following conditions: WCl_6 feeding rate of 3.5 g/min, the flow rate of Ar-H_2-CH_4 mixture to carry the WCl_6 powder of 4 L/min (25°C, 86.1 kPa), applied plasma torch power of 13 kW, and the flow rate of plasma gas (Ar) of 38 L/min (25°C, 86.1 kPa) with no secondary plasma gas (H_2). [Adapted from Ryu et al. (2008a)].

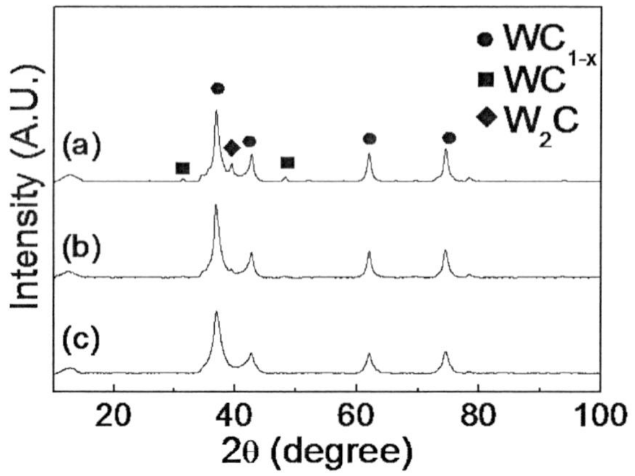

Figure 5.8. X-ray diffraction patterns of the products obtained with different C/W ratios in the feed stream: (a) 6.3, (b) 4.6, (c) 2.5, and (d) 1.5. [Adapted from Ryu et al. (2008a)].

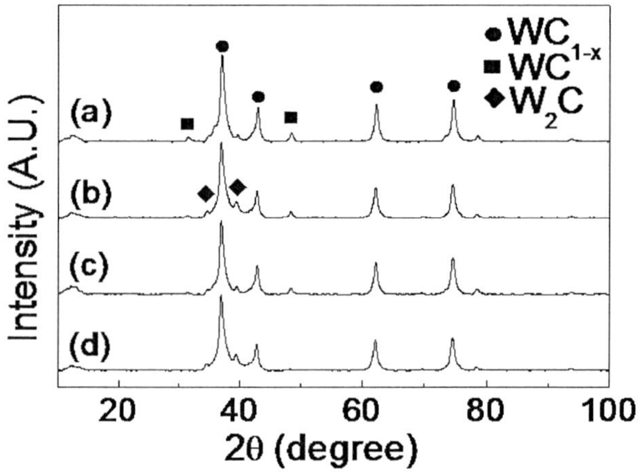

Figure 5.9. X-ray diffraction patterns of the product obtained from (a) 11 kW, (b) 15 kW, (c) 19 kW, (d) 25 kW, and (e) 32 kW under the following conditions: pressure of primary plasma gas (Ar) of 345 kPa, which resulted in a flow rate of 57 L/min (25°C, 86.1 kPa), WCl_6 feeding rate of 3.5 g/min, input C/W molar ratio of 6.3, and flow rate of Ar to carry WCl_6 powder of 1 L/min (25°C, 86.1 kPa). [Adapted from Ryu et al. (2008a)].

Effect of plasma torch power: The effect of plasma torch power was evaluated by varying it from 11 kW to 32 kW while maintaining all other conditions constant. The fixed conditions were: flow rate of primary plasma gas (Ar) of 57 L/min (25°C, 86.1 kPa) in which no secondary plasma gas (H_2) was used, WCl_6 feeding rate of 3.5 g/min, C/W ratio of 6.3, and flow rate of Ar-CH_4 mixture to carry the WCl_6 powder was 2.5 L/min (25°C, 86.1 kPa) with no H_2. CH_4 alone was enough to reduce and carburize tungsten hexachloride to tungsten carbide [Ryu et al., 2008c]. The ratio of H_2 produced from CH_4 to H_2 required to reduce WCl_6 was 4.2 at the C/W ratio of 6.3. The pressure in the reaction chamber was always 86.1 kPa.

Figure 5.9 shows the XRD patterns of the products obtained with different power levels of the plasma torch. The main product was WC_{1-x} and the amounts of WC and W_2C were small in all cases. The grain size of the product WC_{1-x} was 8 ± 1 nm at a power level of 11 kW and increased to 16 ± 1 nm as the applied power was increased to 19 kW or higher, up to 32 kW. The particle size of WC_{1-x} was also examined using TEM. Figure 5.10 shows that the particle size of WC_{1-x} was less than 10 nm at a power level of 11 kW and increased to about 20 nm at an increased plasma torch power of 32 kW. The produced particles were mostly round. The particle size obtained from the TEM micrograph was close to the grain size calculated from the XRD results. Thus, the particles produced by the plasma process were single crystals and it is reasonable to use the grain size calculated from the XRD pattern as the particle size. Since excess methane was used in the reaction, free carbon was also present in the product.

Effect of plasma gas flow rate: The effect of plasma gas flow rate was determined by varying the flow rate of the primary plasma gas (Ar) from 29 L/min (25°C, 86.1 kPa) to 75 L/min (25°C, 86.1 kPa) in which no secondary plasma gas (H_2) was used under otherwise identical conditions [plasma torch power of 13 kW, WCl_6 feeding rate of 3.5 g/min, C/W ratio of 6.3, and flow rate of Ar-CH_4 mixture to carry the WCl_6 powder of 2.5 L/min (25°C, 86.1 kPa) with no H_2]. The main product was WC_{1-x} with a small amount of WC, as shown in Figure 5.11. Although small in all cases, the amount of WC decreased as the plasma gas flow rate was increased. This reduced the time to convert WC_{1-x} to WC. The grain size of WC_{1-x} was 15 ± 1 nm at the plasma gas flow rate of 29 L/min (25°C, 86.1 kPa) and decreased to 9 ± 1 nm when the flow rate was increased to 75 L/min (25°C, 86.1 kPa), as shown in Figure 5.12.

Effect of secondary plasma gas (H_2) addition: The effect of hydrogen addition to the plasma flame was evaluated while keeping all other conditions the same [flow rate of the primary plasma gas (Ar) of 29 L/min (25°C, 86.1 kPa), C/W ratio of 6.3, WCl_6 feeding rate of 3.5 g/min, and flow rate of Ar-CH_4 mixture to carry the WCl_6 powder of 2.5 L/min (25°C, 86.1 kPa) with no H_2]. The addition of secondary plasma gas (H_2) into the plasma flame plays an important role in increasing the plasma flame temperature as well as to enlarge the high temperature flame zone [Choi et al., 2006].

The amount of H_2 added was adjusted to form a composition of Ar-2 mol % H_2 or Ar-7 mol % H_2 in the plasma flame, which required a H_2 flow rate of 0.66 L/min (25°C, 86.1 kPa) or 2.1 L/min (25°C, 86.1 kPa), respectively. The applied power of the plasma torch was 13 kW when no secondary plasma gas (H_2) was added and increased to about 18 kW when the secondary plasma gas (H_2) was added at 0.66 L/min (25°C, 86.1 kPa) or 2.1 L/min (25°C, 86.1 kPa). The predominant phase obtained under these conditions was WC_{1-x} with a small amount of WC, as shown in Figure 5.13. The grain size of WC_{1-x} was 15 ± 1 nm when no secondary plasma gas (H_2) was added and increased to 18 ± 1 nm when the secondary plasma gas (H_2) was added at 0.66 L/min (25°C, 86.1 kPa). However, it decreased to 16 ± 1 nm when the rate of the secondary plasma gas (H_2) was increased to 2.1 L/min (25°C, 86.1 kPa). The addition of the secondary plasma gas (H_2) into the plasma flame promoted further conversion of WC_{1-x} to WC, which resulted from the increased reaction time provided by the expanded flame length. Each of the runs was repeated 2 or 3 times, and thus the variation in grain size and product composition was reproducible. However, no further examination of the reasons for the variation was performed, since the differences were rather small.

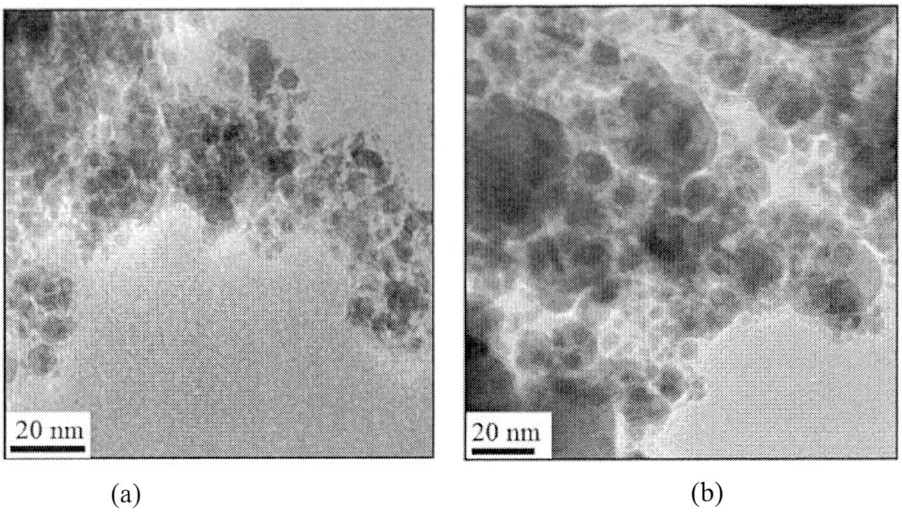

Figure 5.10. TEM photographs of WC_{1-x} nanopowder synthesized at different power levels of plasma torch of (a) 11 kW and (b) 32 kW under the following conditions: flow rate of plasma gas (Ar) of 57 L/min (25°C, 86.1 kPa) with no secondary plasma gas (H_2), WCl_6 feeding rate of 3.5 g/min, C/W ratio of 6.3, and the flow rate of Ar-CH_4 mixture to carry the WCl_6 powder of 2.5 L/min (25°C, 86.1 kPa) with no H_2. [Adapted from Ryu et al. (2008a)].

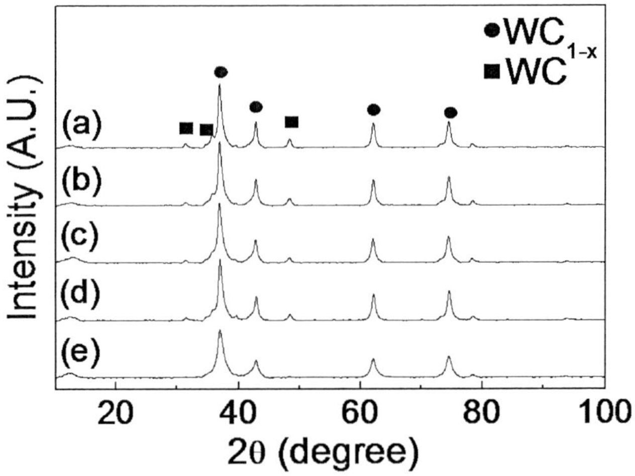

Figure 5.11. X-ray diffraction patterns of the products obtained with various flow rates of primary Ar: (a) 29 L/min (25°C, 86.1 kPa), (b) 38 L/min (25°C, 86.1 kPa), (c) 48 L/min (25°C, 86.1 kPa), (d) 57 L/min (25°C, 86.1 kPa), and (e) 75 L/min (25°C, 86.1 kPa). [Adapted from Ryu et al. (2008a)].

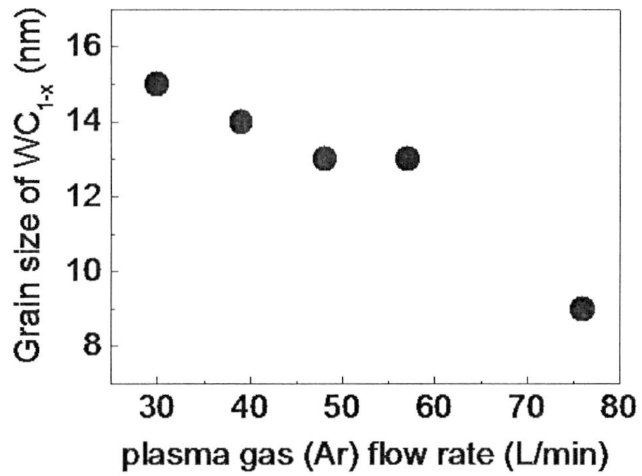

Figure 5.12. Effect of plasma gas flow rate on the grain size of WC_{1-x}. [Adapted from Ryu et al. (2008a)].

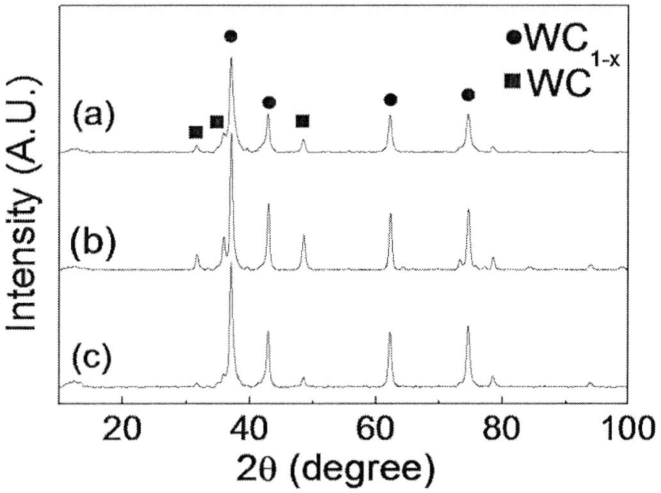

Figure 5.13. X-ray diffraction patterns of the products obtained with various flow rates of the secondary plasma gas (H_2) of (a) 0, (b) 0.66 L/min (25°C, 86.1 kPa), and (c) 2.1 L/min (25°C, 86.1 kPa) under the following conditions: flow rate of the primary plasma gas (Ar) of 29 L/min (25°C, 86.1 kPa), C/W ratio of 6.3, WCl_6 feeding rate of 3.5 g/min, and the flow rate of Ar-CH_4 mixture to carry the WCl_6 powder of 2.5 L/min (25°C, 86.1 kPa) with no H_2. [Adapted from Ryu et al. (2008a)].

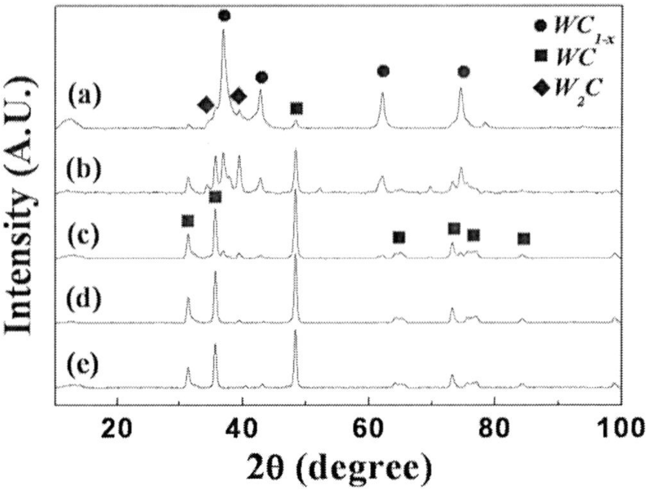

Figure 5.14. X-ray diffraction patterns of the products obtained with different hydrogen treatment times at 900°C using the powder produced from the plasma reactor: (a) before treatment, (b) 1 hour, (c) 2 hours, (d) 3 hours, and (e) 5 hours. [Adapted from Ryu et al. (2008a)].

Removal of free carbon from the produced powder: As mentioned above, free carbon was always present in the produced powder since excess methane was used in the reaction. Thus, the powder obtained from the plasma reactor was subjected to a post-treatment to remove the excess carbon. The powder was placed in a ceramic boat, which in turn was placed in a tube furnace under H_2 atmosphere. The effect of hydrogen heat treatment on the product composition, grain growth, and carbon content was investigated. The amount of powder treated was 10 g, the treatment temperature was 900°C, which was low enough to prevent rapid grain growth of particles. The furnace was heated to 900°C at the rate of 10°C /min and cooled to room temperature also at 10°C /min, and hydrogen was flowed at a rate of 0.5 L/min (25°C, 86.1 kPa) after the temperature reached 900°C. The unreacted WC_{1-x} and W_2C phases were fully carburized to WC phase during this post-treatment, as shown in Figure 5.14. It is likely that the carburization of WC_{1-x} and W_2C phases to fully carburized WC phase occurs at this temperature by highly mobile hydrocarbon, mainly CH_4 formed during the hydrogen post-treatment, which reacts with WC_{1-x} and W_2C particles.

The excess carbon in the product, 213 % before the treatment, was completely removed after the powder was treated for 5 hours, as shown in Figure 5.15. The grain size of WC_{1-x} was 13 ± 1 nm before the treatment and WC powder with grain size of 36 ± 1 nm was obtained after the treatment. Figure 5.16 shows the TEM micrographs of WC_{1-x} powder obtained at 13 kW torch power and the WC powder obtained after the hydrogen heat treatment for 5 hours. The particle size of WC_{1-x} powder was less than about 20 nm before the treatment and that of WC obtained after the treatment was less than about 100 nm. As shown in the TEM micrograph, grain growth and agglomeration of particles were observed after the treatment.

The powder obtained from the plasma reactor was also treated at 800°C, because grain growth occurred during the treatment at 900°C. Figure 5.17 shows the XRD patterns of the products obtained from 2 and 5 hours of hydrogen treatment at 800°C. The main product after the treatment was W_2C with a small amount of WC, which already existed before the treatment. It can be seen that at this temperature WC_{1-x} was decarburized to the W_2C phase rather than being carburized to the WC phase, even though excess carbon was present during the hydrogen treatment.

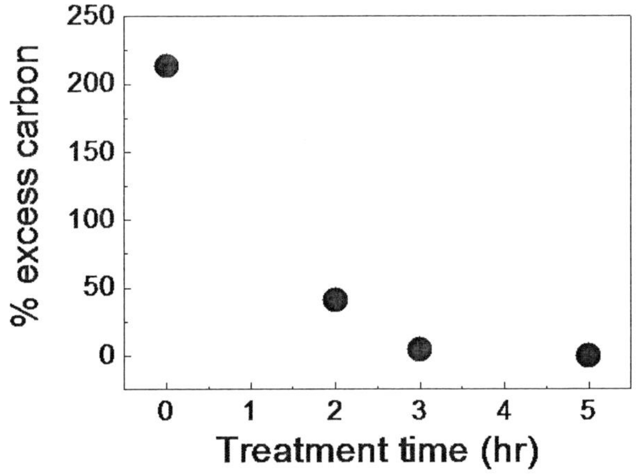

Figure 5.15. Effect of hydrogen heat treatment time at 900°C on the % excess carbon. [Adapted from Ryu et al. (2008a)].

The % excess carbon in the product, defined as the percentage of the excess carbon over the corresponding stoichiometric amount if all W was present as WC, was 213 % before the treatment but decreased to -39 % after the powder was treated for 5 hours at 800°C. From the results, it was concluded that the post-treatment temperature of powders should be 900°C or higher to produce the fully carburized WC phase.

To summarize, nanosized tungsten carbide powder was obtained by a thermal plasma process using tungsten hexachloride (WCl_6) as the precursor. The experimental results showed that nanosized WC_{1-x} powder, which sometimes contained small amounts of WC and W_2C phases, can be synthesized by the thermal plasma process using WCl_6 and CH_4 as the starting materials.

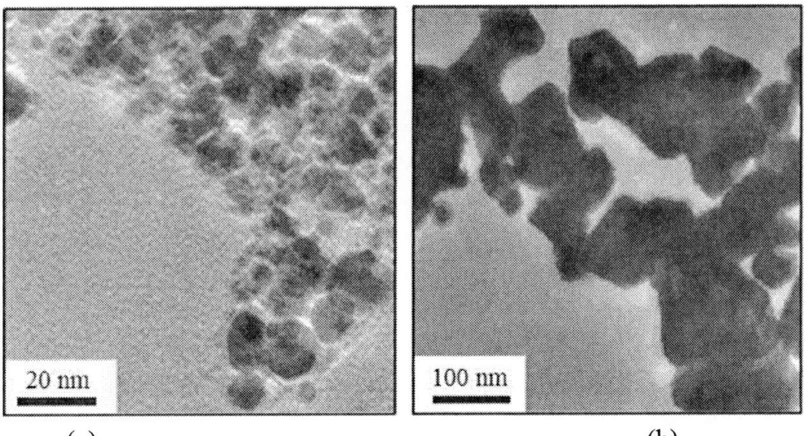

(a) (b)

Figure 5.16. TEM photographs of powders: (a) WC_{1-x} powder obtained from the plasma reactor and (b) WC powder obtained after the hydrogen treatment for 5 hours at 900°C. [Adapted from Ryu et al. (2008a)].

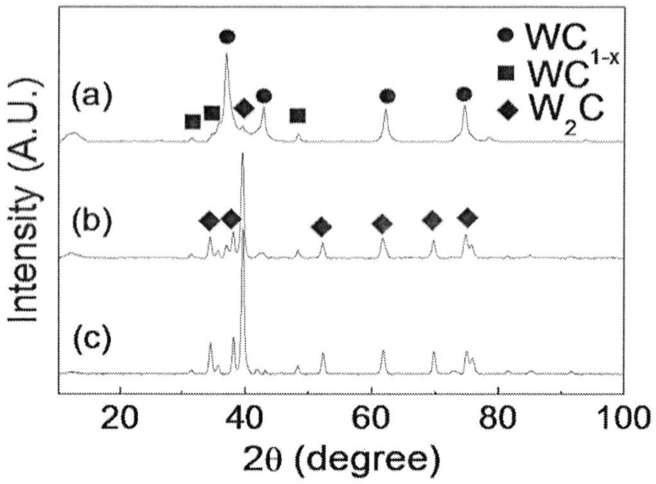

Figure 5.17. X-ray diffraction patterns of the products obtained with different hydrogen heat treatment times at 800°C using the powder obtained from the plasma reactor: (a) before treatment, (b) 2 hours, and (c) 5 hours. [Adapted from Ryu et al. (2008a)].

The amounts of WC and W_2C in the product decreased with increasing plasma torch power and plasma gas flow rate. The particle size of WC_{1-x} was affected by plasma torch power, the plasma gas flow rate, and the addition of secondary plasma gas (H_2) but not by the methane concentration within the range tested. The produced WC_{1-x} and W_2C phases were converted to the fully carburized WC phase and excess carbon in the product was completely removed by hydrogen post-treatment at 900°C. The particle size of WC_{1-x} obtained from the thermal plasma process was less than 20 nm, and WC powder having particle sizes less than 100 nm with grain sizes less than 40 nm was obtained after the hydrogen post-treatment of the produced WC_{1-x} powder. The chemical vapor synthesis conducted in a thermal plasma system has confirmed its feasibility for the preparation of nanosized powders. This process is also suitable for large-scale production of nanosized tungsten carbide powders since the plasma process can work continuously.

From Ammonium Paratungstate (APT)

Tungsten chloride requires more chemical process steps to produce from naturally occurring tungsten minerals than the intermediate product ammonium paratungstate (APT). However, the latter does not yield volatile tungsten compounds at temperatures generated by electrical resistance heating elements. By taking advantage of the extremely high temperatures available in a plasma flame, Ryu et al. [2009b] synthesized nanosized tungsten carbide powder with ammonium paratungstate (APT) as the precursor. The reduction and carburization of vaporized APT by methane-hydrogen mixtures produced nanosized tungsten carbide (WC_{1-x}) powder, which sometimes contained a small amount of W_2C phase. The effects of reactant gas composition, plasma torch power, the flow rate of plasma gas, and the addition of secondary plasma gas (H_2) on the product composition and particle size were investigated. The produced tungsten carbide (WC_{1-x}) particles were less than 20 nm in size. The synthesized powder was subjected to a hydrogen heat treatment to fully carburize the

WC_{1-x} and W_2C phases to the WC phase as well as to remove excess carbon. Finally, WC powder of particle size less than 100 nm was obtained.

The plasma reactor system described in Section 3.2 was used in this work. Ammonium paratungstate (APT, 99.9%, Alldyne Powder Technologies, Huntsville, AL) of an average particle size of 40 μm was used as the precursor. Hydrogen (99.9%, Airgas Intermountain, Salt Lake City, UT) and methane (99.9%, Airgas Intermountain, Salt Lake City, UT) were used as the reducing and carburizing agents, respectively. Argon (99.9%, Airgas Intermountain, Salt Lake City, UT) and hydrogen were used as the primary and the secondary plasma gas, respectively. Argon was also separately passed through the powder feeder as the carrier gas as well as an inert gas to keep the atmosphere in the container inert. A mixture of CH_4 and H_2 also flowed through the precursor feeding system to carry ammonium paratungstate particles into the plasma flame. The produced powder was collected using a Teflon-coated polyester filter with a pore size of 1 μm. The reactor was purged with an Ar flow of 5 L/min (25°C, 86.1 kPa total pressure at Salt Lake City) for 10 minutes before and after each experiment. The product was analyzed by XRD (Siemens D 5000, BRUKER AXS Inc., Madison, WI) and the morphology and particle size analysis was done by TEM (Tecnai F30 TEM, FEI Co., Hillsboro, OR). The total carbon contents of the powders were measured with a Carbon Determinator (LECO CS-444, Elementar Americas, Inc., Mt. Laurel, NJ). The grain size of the synthesized powders was calculated from the XRD pattern by applying the Scherrer equation described in Section 3.2 [Cullity, 1978]. [It is again noted that a particle can be made up of a porous or nonporous polycrystalline solid or just a single crystalline grain.]

Preliminary experiments: Initial experiments were conducted to evaluate the feasibility of APT to be used as the WC precursor without pre-treatment. APT is first decomposed to WO_3 and reduced to W, as indicated earlier in Section 3.2, before being carburized by the following reaction:

$$WO_3 + H_2(g) + CH_4(g) \rightarrow W_2C, WC_{1-x}, WC + H_2O(g) \qquad (5\text{-}1)$$

WC_{1-x} was the major phase in the product, as explained in the previous section based on the phase diagram available in the literature [Sara, 1965]. The grain sizes of WC_{1-x} powders were less than 15 nm in all cases tested. The shapes of WC_{1-x} particles examined by TEM were mostly round. In the subsequent experiments, other factors that affect the WC_{1-x} formation were further examined, as described below.

Effect of methane concentration: The effect of methane concentration was evaluated by varying the C/W ratio in the feed stream. The experimental conditions were as follows: The feeding rate of APT was 3.1 g/min, the applied power of the plasma torch was 16 kW, the flow rate of the plasma gas (Ar) was 38 L/min (25°C, 86.1 kPa) with no secondary plasma gas (H_2), the flow rate of Ar-H_2-CH_4 mixture to carry the APT powder was 4 L/min (25°C, 86.1 kPa), and the C/W ratio in the feed stream was varied from 3.7 to 5.7. The main product under these conditions was WC_{1-x}, which contained a small amount of W_2C phase, as shown in Figure 5.18. The amount of W_2C decreased as the C/W ratio in the feed stream was increased. The grain size of WC_{1-x} at 11 ± 1 nm was not affected by the methane concentration in the feed stream within the range tested. The weight percent of carbon in the product was converted to percent (%) excess carbon, which is defined as the percentage of the excess carbon over the corresponding stoichiometric amount if all W was present as WC. The % excess carbon in the product obtained at a C/W ratio of 5.7 in the feed stream was 130 %.

It decreased to 29 % as the C/W ratio was decreased to 3.7. This level of free carbon can be tolerated since it can be removed by post-treatment using hydrogen, as will be discussed later. The subsequent experiments were performed at a C/W ratio of 5.7 which produced a lower amount of W_2C.

Effect of plasma torch power: The effect of plasma torch power was determined by varying it from 7 kW to 29 kW under otherwise identical conditions [flow rate of primary plasma gas of 38 L/min (25°C, 86.1 kPa) without secondary plasma gas (H_2) addition, APT feeding rate of 3.1 g/min, C/W ratio of 5.7 in the feed stream, and the flow rate of CH_4-H_2 mixture to carry APT powder of 4 L/min (25°C, 86.1 kPa) in which no Ar was used in the carrier gas]. Figure 5.19 shows the XRD patterns of the products, indicating that they contained WC_{1-x} with a small amount of W_2C phase. The product composition was not affected by plasma torch power but the grain size of WC_{1-x} increased as the plasma torch power was increased within the range tested.

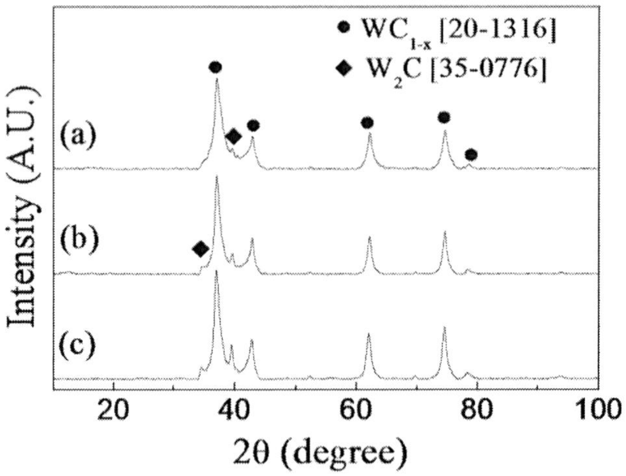

Figure 5.18. X-ray diffraction patterns of the products obtained with various C/W ratios in the feed stream: (a) 5.7, (b) 4.7, and (c) 3.7. [The numbers after the phases are the JCPDS file numbers.] [Adapted from Ryu et al. (2009b)].

The grain size of the product WC_{1-x} was 5 ± 1 nm at 7 kW of plasma torch power and increased to 13 ± 1 nm as the plasma torch power was increased to 29 kW. The particle sizes of WC_{1-x} were also examined using TEM micrographs. Figure 5.20 shows that the particle size of WC_{1-x} was less than 10 nm at the power level of 7 kW and increased to an average size of about 15 nm with the spread ranging from 10 nm to 40 nm as the plasma torch power was increased to 29 kW. The particle size obtained from TEM micrograph was close to the grain size calculated from the XRD pattern. Thus, the particles produced by the plasma process were single crystals and it is reasonable to use the calculated grain size from XRD result as the particle size.

Effect of plasma gas flow rate: The effect of plasma gas (Ar) flow rate was investigated by varying it from 29 L/min (25°C, 86.1 kPa) to 66 L/min (25°C, 86.1 kPa) without secondary plasma gas (H_2) addition while keeping all other conditions the same [plasma torch power of 16 kW, APT feeding rate of 3.1 g/min, C/W ratio of 5.7 in the feed stream, and the flow rate of CH_4-H_2 mixture to carry APT powder of 4 L/min (25°C, 86.1 kPa) in which no Ar was used in the carrier gas].

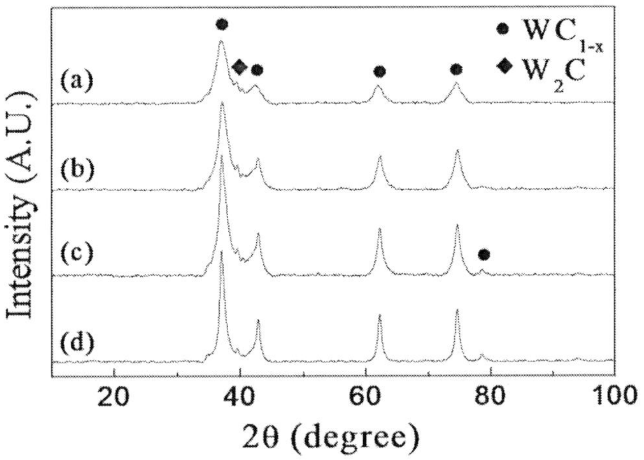

Figure 5.19. X-ray diffraction patterns of the products obtained with different power levels of plasma torch: (a) 7 kW, (b) 10 kW, (c) 16 kW, and (d) 29 kW under the following conditions: flow rate of plasma gas of 38 L/min (25°C, 86.1 kPa) without secondary plasma gas (H_2) addition, APT feeding rate of 3.1 g/min, C/W ratio of 5.7 in the feed stream, and the flow rate of CH_4-H_2 mixture to carry APT powder of 4 L/min (25°C, 86.1 kPa) with no Ar was used as a carrier gas. [Adapted from Ryu (2008)].

The product was WC_{1-x} and a small amount of W_2C phase was present in the product, as shown in Figure 5.21. Although small in all cases, the amount of W_2C slightly increased as the flow rate of plasma gas was increased. The grain size of WC_{1-x} was 11 ± 1 nm at the plasma gas flow rate of 29 L/min (25°C, 86.1 kPa) and was not affected by the flow rate of plasma gas within the range tested. It was expected that WC_{1-x} grain size would decrease with increasing plasma gas flow rate, which results in the shorter residence time of produced particles. However, the change of grain size observed in the results was not noticeable. Each of the runs was repeated 2 or 3 times, the results of which indicated that the average grain size (within 1 nm) and product composition were reproducible.

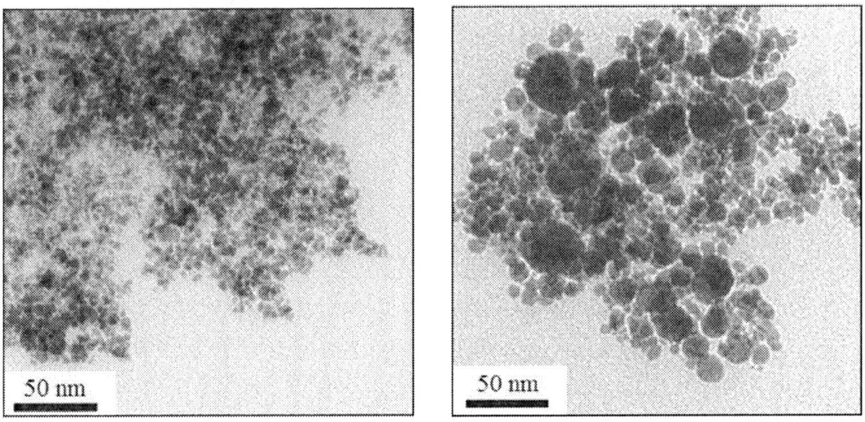

(a) (b)

Figure 5.20. TEM micrographs of WC_{1-x} nanopowders synthesized at various power levels of plasma torch: (a) 7 kW and (b) 29 kW. [Adapted from Ryu et al. (2009b)].

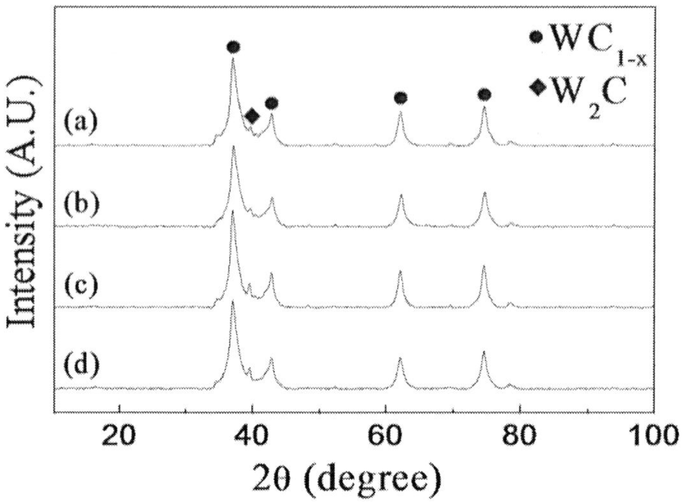

Figure 5.21. X-ray diffraction patterns of the products obtained with various plasma gas flow rates: (a) 29 L/min (25°C, 86.1 kPa), (b) 38 L/min (25°C, 86.1 kPa), (c) 48 L/min (25°C, 86.1 kPa), and (d) 66 L/min (25°C, 86.1 kPa). [Adapted from Ryu (2008)].

Effect of secondary plasma gas (H_2) addition: Experiments were conducted to evaluate the effect of adding hydrogen in the plasma flame while keeping all other conditions the same [flow rate of the primary plasma gas (Ar) of 38 L/min (25°C, 86.1 kPa), C/W ratio of 5.7 in the feed stream, APT feeding rate of 3.1 g/min, and the flow rate of CH_4-H_2 mixture to carry APT powder of 4 L/min (25°C, 86.1 kPa) in which no Ar was used in the carrier gas]. The addition of secondary plasma gas (H_2) into the plasma flame plays an important role in increasing the plasma flame temperature as well as enlarging the high temperature flame zone [Choi et al., 2006]. The amount of H_2 added was adjusted to form a composition of Ar-0.14 mol % H_2, Ar-1.7 mol % H_2, or Ar-4.6 mol % H_2 in the plasma flame, which required a H_2 addition rate of 0.05 L/min (25°C, 86.1 kPa), 0.67 L/min (25°C, 86.1 kPa), or 1.89 L/min (25°C, 86.1 kPa), respectively. The applied power of the plasma torch was 16 kW when no secondary plasma gas (H_2) was added and it increased to about 22 kW when the secondary plasma gas (H_2) was added regardless of its flow rate. The predominant phase in the product obtained under these conditions was WC_{1-x} with a small amount of W_2C, as shown in Figure 5.22. Even though the addition of secondary plasma gas (H_2) into the plasma flame increased the plasma flame temperature and enlarged high temperature region, the rapid quenching of produced particles still resulted in the formation of mostly WC_{1-x} phase. In addition, the increase of hydrogen concentration in the reaction inhibited the decomposition of CH_4 to C and H_2, which resulted in a decreased carbon content. Thus, the amount of W_2C phase, although small, increased compared with the case in which no secondary plasma gas (H_2) was added. The % excess carbon, 130 % without the addition of the secondary plasma gas (H_2), decreased to 13 % with the addition of secondary plasma gas (H_2) of 1.89 L/min (25°C, 86.1 kPa). The % excess carbon in the products obtained with lower flow rates of the secondary gas was not analyzed. The grain size of WC_{1-x} was 11 ± 1 nm when no secondary plasma gas (H_2) was added and increased to 14 ± 1 nm when the secondary plasma gas (H_2) was added at 1.89 L/min (25°C, 86.1 kPa).

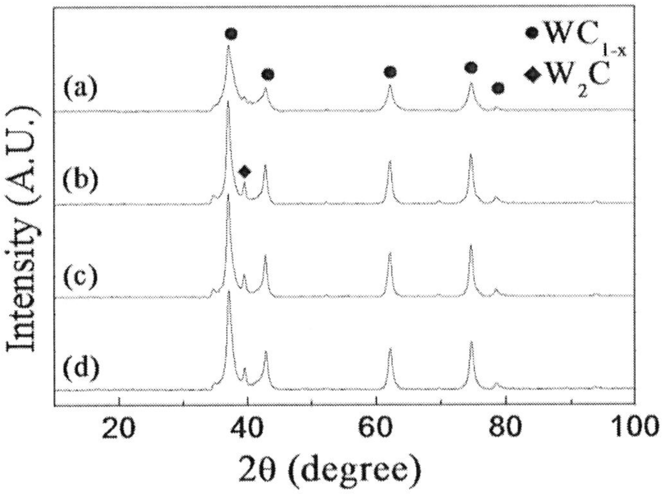

Figure 5.22. X-ray diffraction patterns of the products obtained by the addition of secondary plasma gas (H_2) into the plasma flame at various rates: (a) 0, (b) 0.05 L/min (25°C, 86.1 kPa), (c) 0.67 (25°C, 86.1 kPa), and (d) 1.89 L/min (25°C, 86.1 kPa). [Adapted from Ryu (2008)].

Post-treatment of the produced powder: The amount of free carbon decreased with decreasing methane concentration in the feed stream and with the addition of the secondary plasma gas (H_2) into the plasma flame. However, free carbon was still present in the product, while these changes resulted in increased formation of the W_2C phase in the product. The powder obtained from the plasma reactor was placed in a ceramic boat of 3.6 cm W x 9 cm L x 1.5 cm H, which in turn was placed in a tube furnace under an H_2 atmosphere. The effect of hydrogen heat treatment on the product composition, grain growth, and carbon content were investigated. The amount of powder treated was 5 g, the treatment temperature was kept relatively low at 900°C, the heating and cooling rates were 10°C/min, and hydrogen was flowed at a rate of 0.5 L/min (25°C, 86.1 kPa) after the temperature reached 900°C. This temperature was low enough to prevent rapid grain growth of particles but high enough to complete post-treatment reactions within a reasonable length of time. The % excess carbon in the product, 130 % before the treatment, was down to 35 % after 1 hour of post-treatment, 14 % after 2 hours, and 0 % after 3 hours. The unreacted WC_{1-x} and W_2C phases were fully carburized to the WC phase after the powder was treated by hydrogen for 3 hours, as shown in Figure 5.23.

The grain size of WC_{1-x} was 11 ± 1 nm before the treatment, and WC powder with an average grain size of 43 ± 1 nm was obtained after the treatment for 3 hours. Figure 5.24 shows the TEM micrographs of WC_{1-x} powder obtained from the plasma reactor at the plasma torch power of 16 kW and the WC powder obtained after the hydrogen heat treatment for 3 hours. The particle size of WC_{1-x} was less than 15 nm before the treatment and that of the WC obtained after the treatment was less than 100 nm.

The post-treatment was also tested at 800°C to suppress the particle growth observed at 900°C. However, at this temperature the decarburization of WC_{1-x} phase to W_2C phase was dominant rather than carburization to WC phase, even though excess carbon was present during the hydrogen treatment [Ryu et al., 2008a].

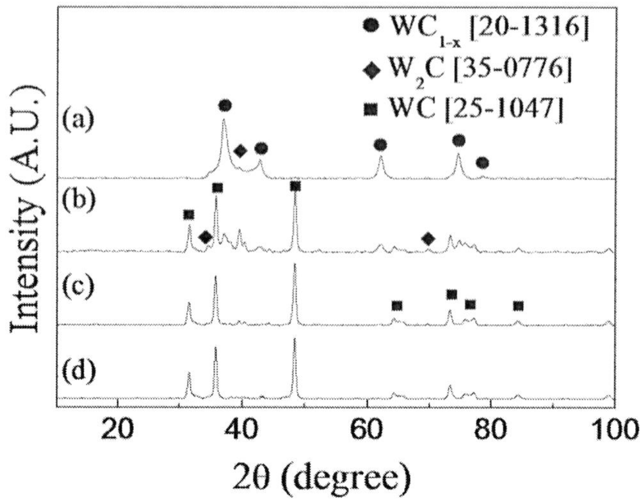

Figure 5.23. X-ray diffraction patterns of the product obtained with various hydrogen treatment times at 900°C using powders produced from the plasma reactor: (a) before treatment, (b) 1 hour, (c) 2 hours, and (d) 3 hours. [The numbers after the phases are the JCPDS file numbers.] [Adapted from Ryu et al. (2009b)].

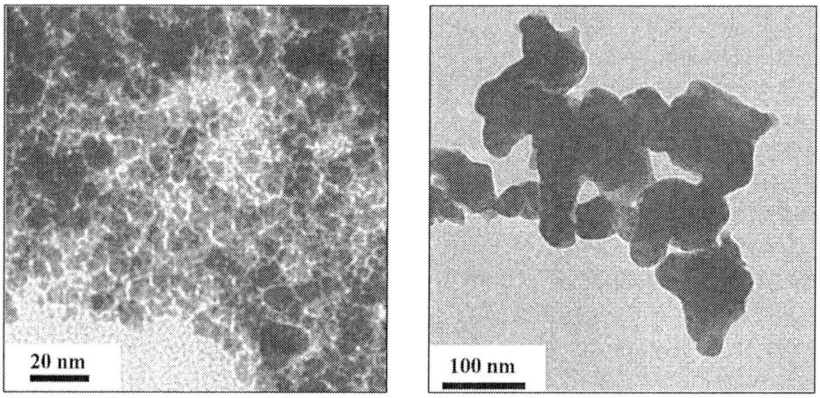

(a) (b)

Figure 5.24. TEM micrographs: (a) WC_{1-x} powder obtained from the plasma reactor and (b) WC powder obtained after the hydrogen heat treatment of WC_{1-x} powder for 3 hours at 900°C. [Adapted from Ryu et al. (2009b)].

To summarize, nanosized tungsten carbide (WC_{1-x}) powder was synthesized by a thermal plasma process using ammonium paratungstate as the starting material. The particle size of synthesized WC_{1-x} was affected by plasma torch power and the addition of secondary plasma gas (H_2). The product composition was not affected by plasma torch power. The amount of W_2C phase in the product, although small in all cases tested, increased with decreasing methane concentration in the feed stream, increasing flow rate of plasma gas, and the addition of secondary plasma gas (H_2). The amount of free carbon in the product decreased with decreasing methane concentration in the feed stream and with the addition of secondary plasma gas (H_2). A summary of the experimental results obtained with various processing parameters tested is listed in Table 5-1. The synthesized WC_{1-x} powder containing a small amount of W_2C phase was converted to the fully carburized WC phase and excess carbon in the product was removed by post-treatment using hydrogen. The particle size of WC_{1-x} synthesized from the thermal plasma process was less than about 20 nm and WC powder with a particle size less than about 100 nm was obtained after the post-treatment under the conditions tested.

Table 5.1. Summary of experimental results

Test parameter	Range	Effect	Other fixed conditions
Methane concentration	C/W ratio: 3.7 ~ 5.7	- Increased W_2C phase with decreasing C/W ratio - Grain size of WC_{1-x}, 11 ± 1 nm, not affected by the methane concentration - % excess carbon, 130 % at C/W ratio of 5.7 decreased to 29 % at C/W ratio of 3.7	- APT feeding rate = 3.1 g/min - Plasma torch power = 16 kW - Plasma gas flow rate = 38 L/min (25°C, 86.1 kPa) - Flow rate of carrier gas (Ar-CH_4-H_2) = 4 L/min (25°C, 86.1 kPa)
Plasma torch power	7 ~ 29 kW	- The product composition, not affected - Grain size of WC_{1-x}, 5 ± 1 nm at 7 kW increased to 13 ± 1 nm at an increased plasma torch power of 29 kW	- APT feeding rate = 3.1 g/min - C/W ratio = 5.7 - Plasma gas flow rate = 38 L/min (25°C, 86.1 kPa) - Flow rate of carrier gas (CH_4-H_2) = 4 L/min (25°C, 86.1 kPa)
Plasma gas (Ar) flow rate	29 ~ 66 L/min (25°C, 86.1 kPa)	- Slightly increased W_2C phase in the product with increasing flow rate - Grain size of WC_{1-x}, 11 ± 1 nm, not affected by the flow rate	- APT feeding rate = 3.1 g/min - C/W ratio = 5.7 - Plasma torch power = 16 kW - Flow rate of carrier gas (CH_4-H_2) = 4 L/min (25°C, 86.1 kPa)
Addition of secondary plasma gas (H_2)	0.05 ~ 1.89 L/min (25°C, 86.1 kPa)	- Slightly increased W_2C phase in the product with the addition of secondary plasma gas (H_2) - % excess carbon, 130 % without the addition, decreased to 13 % with the addition of secondary plasma gas (H_2) of 1.89 L/min (25°C, 86.1 kPa) - Grain size of WC_{1-x}, 11 ± 1 nm with no secondary plasma gas (H_2) increased to 14 ± 1 nm when the secondary plasma gas (H_2) was added at 1.89 L/min (25°C, 86.1 kPa)	- APT feeding rate = 3.1 g/min - C/W ratio = 5.7 - Plasma gas flow rate = 38 L/min (25°C, 86.1 kPa) - Flow rate of carrier gas (CH_4-H_2) = 4 L/min (25°C, 86.1 kPa)

The thermal plasma process performed in this work confirmed its potential as an efficient technique compared with other conventional multi-step processes on the preparation of nanosized tungsten carbide, owing to the high temperature to rapidly volatilize APT powder, followed by vapor phase reaction and rapid quenching of the product to directly yield nanosized powder.

5.3. Synthesis from Tungsten Carbonyl

Kim and Kim [2004] used tungsten hexacarbonyl [$W(CO)_6$] as the precursor and carbon monoxide (CO) as the carburizing agent. The experiments were conducted in the temperature range of 500°C to 1000°C, and three different reaction atmospheres, 1 atm air, 1 atm argon, and 1×10^{-4} atm vacuum, were tested. The solid precursor was kept at 120°C to vaporize it in the vaporizer and no H_2 was used as a reducing agent in this experiment since tungsten hexacarbonyl is decomposed into W and CO, typically in the temperature range from 150°C to 205°C. CO undergoes a reaction to form CO_2 and C in the presence of tungsten, which then carburizes the W to form WC_{1-x}. No powder was obtained under 600°C, and the product composition, mainly WC_{1-x}, obtained above this temperature, remained almost the same at different reaction temperatures. The particle size increased with reaction temperature. The synthesized powders in air and Ar atmospheres had the same XRD patterns and the XRD peaks were greatly broadened under vacuum condition compared with those under air and Ar atmospheres. The particle sizes obtained in these experiments were less than 50 nm.

Chapter 6

TUNGSTEN CARBIDE – COBALT COMPOSITE NANOPOWDER

ABSTRACT

As mentioned earlier, the chemical vapor synthesis (CVS) process is ideally suited for producing very uniformly mixed particles of distinct solid phases. This chapter describes the synthesis of WC-Co composite powder in a single reactor.

As mentioned in the previous section, cemented tungsten carbide has a wide range of applications in such areas as metal working, drilling and mining industries under severe conditions of high pressure, and high temperature and corrosive environments thanks to its outstanding properties of high hardness and good wear resistance. However, its performance has been limited by brittleness and relatively low toughness. Cobalt is typically added as a ductile matrix to improve the fracture toughness, which is a function of microstructural variables including metal contents, grain size and contiguity of tungsten carbide (WC) grains. The mechanical properties such as hardness, compressive strength and transverse rupture strength depend on the composition and microstructural parameters such as the cobalt contents and the grain size of WC [Upadhaya, 2002; Wahlberg et al. 1997; Zhu and Manthiram 1996]. Previous investigations [Zawrah, 2007; Lee and Kang, 2006; Petersson and Ågren, 2004; Wu et al., 2004; Fang and Eason, 1995] have shown that the reduction of tungsten carbide grain size gives significant improvements of the mechanical properties. Cobalt pooling during consolidation process is effectively reduced when the particle size of cobalt is small [Krakhmalev et al., 2007; Carroll, 1999; Kear and McCandlish, 1993]. Cobalt contents in WC-Co composites are typically in the range of 3 - 30 wt % for commercial applications.

6.1. SYNTHESIS FROM CHLORIDES

Sohn and coworkers [Sohn et al., 2007; Ryu et al.; 2008c; Ryu et al.; 2008d] prepared nanograined WC-Co composite powders by the simultaneous reduction and carburization of WCl_6 and $CoCl_2$ vapors by a gas mixture containing CH_4 and H_2. The reaction produced WC and Co composite powder, which sometimes contained W_2C, W and/or the η (Co_3W_3C) phase.

In this work, WCl_6 (99.9%) and $CoCl_2$ (99.7%) were used as the precursors, and hydrogen (99.9%) and methane (99.9%) were used, respectively, as the reducing and carburizing agents. The overall reactions for synthesizing the WC-Co mixtures using metal chloride powders are as follows:

$$WCl_6(g) + xCoCl_2(g) + CH_4(g) + (x+1)H_2(g) = WC(s) + xCo(s) + (2x+6)HCl(g) \quad (6\text{-}1)$$

The equilibrium composition diagram for the WCl_6-$CoCl_2$-H_2-CH_4 mixture was obtained by the use of HSC Chemistry software developed by Outotec [2002], as shown in Figure 6.1. The diagram shows it is thermodynamically feasible to synthesize WC-Co composites from the WCl_6-$CoCl_2$-H_2-CH_4 mixture.

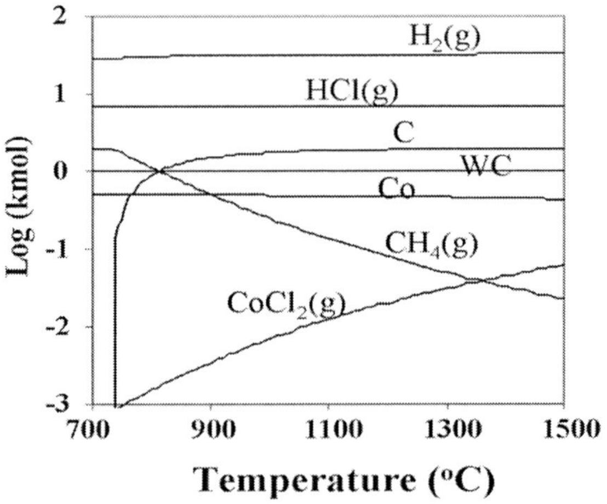

Figure 6.1. Calculated equilibrium composition in WCl_6-$CoCl_2$-H_2-CH_4 system with input amounts of WCl_6, $CoCl_2$, H_2 and CH_4 of 1, 0.5, 30, and 3 kmol, respectively. [Adapted from Sohn and Ryu, 2010].

Experimental Apparatus and Procedure

The same apparatus used for the synthesis of WC powder described in Chapter 5 was used in this work. One of the two entrained-flow powder feeders was used to feed $CoCl_2$ powder. Argon was flowed to carry the $CoCl_2$ powder into a $CoCl_2$ volatilizer to avoid the deposition of reduced Co in the volatilizer. Because it had been observed during earlier experiments that an intermediate phase Co_3W_3C was formed when premixed powders of WCl_6 and $CoCl_2$ were fed into the reactor [Ryu et al., 2008c; 2008d], the chlorides were fed separately. Furthermore, the WCl_6 volatilizer temperature was sufficiently low to prevent the reduction and/or carburization of WCl_6 by methane in the volatilizer, which would result in a deposition of solids in the volatilizer. Hojo et al. [1978] reported that substantial WC formation from WCl_6 starts above 1000~1200°C, as also observed in the current work. The formation of an intermediate phase Co_3W_3C or Co_6W_6C was avoided by placing the $CoCl_2$ volatilizer at a point where temperature was 1400°C. Thus, tungsten was given a sufficient length of time to be carburized by the time it is carried to this position of $CoCl_2$ volatilizer. The deposition of reduced Co in the $CoCl_2$ volatilizer was prevented by using Ar to carry $CoCl_2$ powder so that it can be reduced by methane away from the volatilizer.

The product was analyzed by XRD (Siemens, D 5000) and the morphology was examined by TEM (JEOL, JEM-2000FXII) with energy dispersive X-ray spectrometry (EDS) that also gave distribution of the elements. The total carbon contents of the powders were measured with a Carbon Determinator (LECO, CS-444). The grain size of WC was calculated from the XRD pattern by applying the Scherrer equation calibrated by using standard samples [Cullity, 1978], as described in Section 3.2. It is noted that because the powders in this study were produced in the vapor phase, no lattice strain effects were considered in this grain size analysis. The average particle size, as opposed to the grain size, was measured by using a Brookhaven Instruments ZetaPALS unit, which is a zeta potential analyzer based on phase analysis light scattering.

The molar ratio of the different tungsten carbide phases in the product was estimated from the XRD patterns by applying the internal standard method, combined with the following calibration equation [Cullity, 1978];

$$\frac{I_\alpha}{I_\beta} = \kappa \frac{X_\alpha}{X_\beta} \tag{6-2}$$

where I_α = integrated intensity of α phase peak; I_β = integrated intensity of β phase peak; X_α = mole fraction of α phase; X_β = mole fraction of β phase; and κ = constant.

The cobalt content in WC-Co composite powder can be adjusted by controlling the feeding rate of WCl_6 or $CoCl_2$ powder in the feed stream. In this work the Co/W ratio in the composite product was 0.8, which is equivalent to the cobalt content of 19 wt % without free carbon after the post-treatment of the powder as described subsequently.

Experimental Results

In preliminary experiments, a mixture of WCl_6 and $CoCl_2$ powders was fed to produce WC-20 wt% Co mixture. The volatilizer in which the reactant precursors were vaporized was placed at a position where the temperature was 1100°C, considering the boiling point of $CoCl_2$. The main product under this condition was W_2C and even Co_3W_3C was produced as shown in Figure 6.2(a). Thus, a modification was made to the reactant feeding arrangement [Ryu et al., 2008c; 2008d]. In the modified system, separate feeders and volatilizers for WCl_6 and $CoCl_2$ were used to prevent the unreacted precursor gases from coming into contact with each other. The volatilizer for WCl_6 and $CoCl_2$ were placed where the temperatures were 440°C and 1400°C, respectively. The degree of carburization was increased and the formation of Co_3W_3C was suppressed with this modification, and the main product was a mixture of WC and Co with some free carbon as shown in Figure 6.2(b). The grain size of WC was 24 ± 1 nm which was somewhat larger than that of WC synthesized in the WCl_6-CH_4-H_2 system without $CoCl_2$. The presence of small quantities of incompletely carburized phases such as W_2C and η-phase (Co_3W_3C) can be tolerated because they can be fully carburized during the subsequent sintering process as mentioned above. These phases can also be fully carburized by a separate post-treatment. Free carbon was always present in the product because CH_4 was used in excess.

Effect of CH_4 concentration: The effect of CH_4 concentration was investigated by varying its feeding rate from 0.1 L/min (25°C, 86.1 kPa) to 0.01 L/min (25°C, 86.1 kPa) while keeping all other conditions the same (reaction temperature of 1400°C, total flow rate of 1.1 L/min (25°C, 86.1 kPa), WCl_6 feeding rate of 0.06 g/min, and $CoCl_2$ feeding rate of 0.02 g/min). At the lowest feeding rate of CH_4, the C/W (CH_4/WCl_6) molar ratio was 2.3 and the ratio of H_2 produced from CH_4 to H_2 required to reduce both precursors was 1.2.

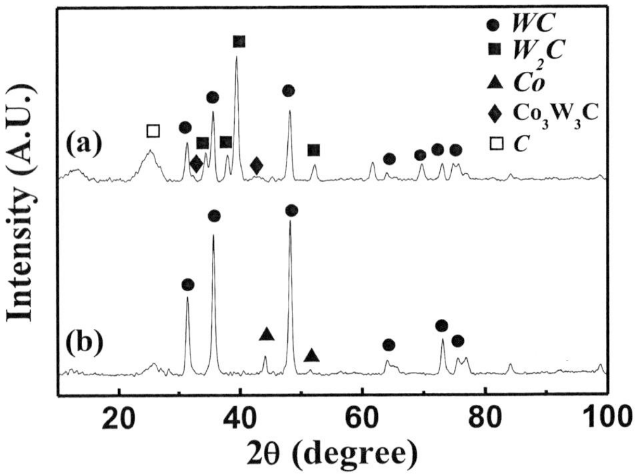

Figure 6.2. X-ray diffraction patterns of the product obtained with (a) mixture feeding ($WCl_6/CoCl_2$ molar ratio of 1.1, and feeding rate of mixture of 0.053 g/min) and (b) modified arrangement (WCl_6 feeding rate of 0.06 g/min, and $CoCl_2$ feeding rate of 0.02 g/min) under the following conditions: reaction temperature = 1400°C, CH_4 feeding rate = 0.1 L/min (25°C, 86.1 kPa), H_2 feeding rate = 0 L/min, and total flow rate = 0.6 L/min (25°C, 86.1 kPa). [Adapted from Ryu et al. (2008c)].

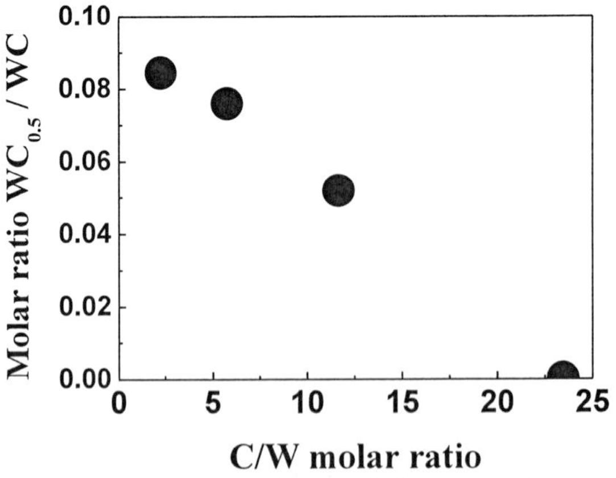

Figure 6.3. Effect of the CH_4/WCl_6 molar ratio on the $WC_{0.5}/WC$ molar ratio. [Adapted from Ryu et al. (2008c)].

Figure 6.3 shows the relationship between $WC_{0.5}$/WC molar ratio and C/W molar ratio. The $WC_{0.5}$/WC ratio is the same as twice the W_2C/WC ratio but better represents how much of W is in each carbide phase. The ratio was calculated from the intensity of the W_2C peak at 2θ of 39.5 and that of the WC peak at 2θ of 35.6, according to Equation (2). This result shows that the degree of carburization increased with increasing carbon potential. The main product was still WC and Co with a small amount of W_2C even at the lowest C/W molar ratio of 2.3. In addition, the amount of free carbon also decreased as the C/W molar ratio decreased.

The average grain size of WC produced under these conditions was 25 ± 1 nm, which was not affected by the change of carbon potential. The particle size of the product obtained at the C/W ratio of 2.3 was determined using ZetaPALS and TEM. ZetaPALS gave an average value of 380 nm, but examination using TEM showed that the actual particle size was 25 ± 5 nm. Figure 6.4 shows the TEM photograph of the WC and Co particles. The much larger particle size obtained by ZetaPALS can be explained by the agglomeration of the synthesized powder as well as motion of the particles in liquid media that contributed to errors in particle size measurement. In addition, as shown in the TEM photograph, excess carbon was coated on the surface of produced particles. The TEM examination of powders showed the same particle size as the grain size calculated by the Scherrer equation. Thus, the produced particles are single crystals and it is reasonable to use the calculated grain size from XRD result as the particle size.

The uniformity of mixing of the WC and Co nanoparticles can be seen from the EDX mapping of the powder shown in Figure 6.5. This presents another example of the significant feature of the chemical vapor synthesis (CVS) process to produce very uniformly mixed powders.

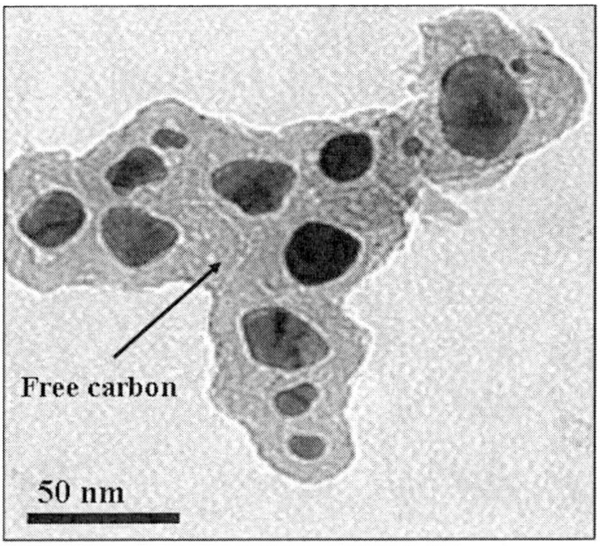

Figure 6.4. TEM photograph of WC-Co nanopowder synthesized under the following conditions: reaction temperature = 1400°C, total flow rate = 1.1 L/min (25°C, 86.1 kPa), CH_4 feeding rate = 0.01 L/min (25°C, 86.1 kPa), no H_2, WCl_6 feeding rate = 0.06 g/min, and $CoCl_2$ feeding rate = 0.02 g/min. [Adapted from Ryu et al. (2008c)].

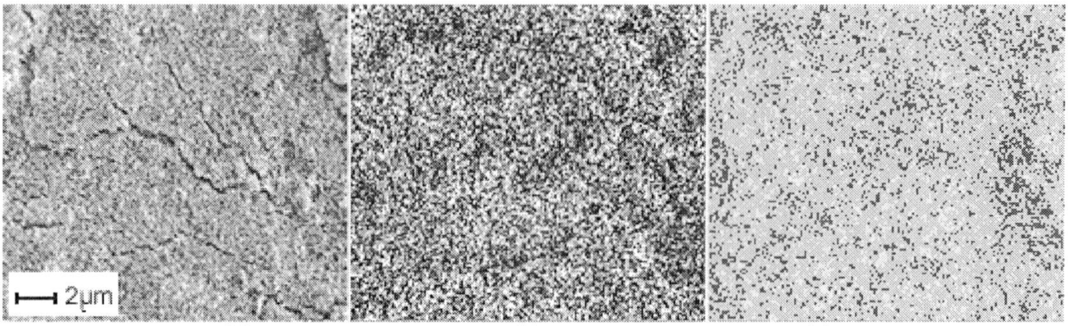

Figure 6.5. SEM image of WC-Co composite powder with associated EDX maps for W and Co: (a) WC-Co composite powder, (b) EDX map of W (light area), (c) EDX map of Co (gray area) under the synthesis conditions: reaction temperature = 1400°C, WCl_6 feeding rate = 0.6 g/min, $CoCl_2$ feeding rate = 0.2 g/min, C/W ratio = 2.3 in the feed stream, no H_2, and total flow rate = 1.1 L/min (25°C, 86.1 kPa). [Adapted from Sohn et al. (2007)].

Effect of H_2 feeding rate: Figure 6.6 shows the XRD patterns of the product obtained at three different H_2 flow rates, 0, 0.01, and 0.05 L/min (25°C, 86.1 kPa), under otherwise identical conditions [reaction temperature of 1400°C, total flow rate of 1.1 L/min (25°C, 86.1 kPa), CH_4 feeding rate of 0.01 L/min (25°C, 86.1 kPa), WCl_6 feeding rate of 0.06 g/min, and $CoCl_2$ feeding rate of 0.02 g/min].

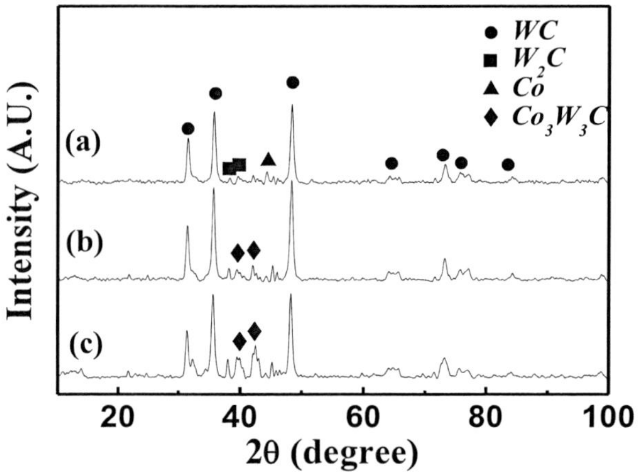

Figure 6.6. X-ray diffraction patterns of the product obtained at various H_2/CH_4 molar ratios (a) H_2/CH_4 = 0, (b) H_2/CH_4 = 1, and (c) H_2/CH_4 = 5 under the following conditions: reaction temperature = 1400°C, total flow rate = 1.1 L/min (25°C, 86.1 kPa), CH_4 feeding rate = 0.01 L/min (25°C, 86.1 kPa), WCl_6 feeding rate = 0.06 g/min, and $CoCl_2$ feeding rate = 0.02 g/min. [Adapted from Ryu et al. (2008c)].

Hydrogen was added to the reactant gas to reduce the amount of free carbon, and the effect of hydrogen concentration in the feed stream was investigated by controlling the H_2/CH_4 ratio from 0 to 4.3 under otherwise identical conditions [C/W ratio of 2.7, reaction temperature of 1400°C, total gas flow rate of 1.1 L/min (25°C, 86.1 kPa), WCl_6 feeding rate of 0.6 g/min, and $CoCl_2$ feeding rate of 0.2 g/min]. The variation in the mole fractions of tungsten carbides obtained with different H_2/CH_4 ratios, shown in Figure 6.7, indicates that the degree of carburization decreased as the H_2/CH_4 ratio increased. It is likely that hydrogen addition suppressed the decomposition of CH_4 into C and H_2. Thus, the C potential decreased as the feeding rate of H_2 increased, resulting in a lower degree of carburization.

The weight percentage of carbon in the product was converted to percent excess carbon, which is defined as the percentage of the excess carbon over the corresponding stoichiometric amount if all W was present as WC. The % excess carbon in the product decreased from 67.8 % to -17.5 % as the H_2/CH_4 ratio in the feed stream increased from 0 to 4.3, as shown in Figure 6.8. The grain size of produced WC decreased from 30 ± 1 nm with no H_2 to 26 ± 1 nm with H_2/CH_4 = 4.3.

Tungsten carbide powder is used mainly to make bulk components by consolidation process. Therefore, the presence of some incompletely carburized W_2C and W phases can be tolerated because they can be fully carburized during the subsequent sintering process. These phases can also be fully carburized during the separate post-treatment of the produced composite powder as discussed subsequently [Ryu et al., 2008d].

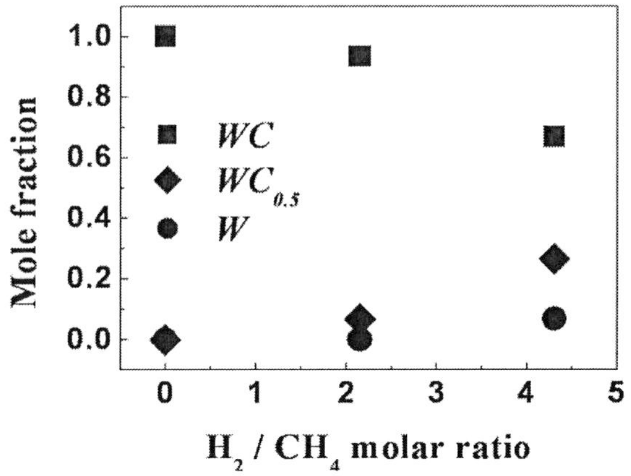

Figure 6.7. Effect of H_2/CH_4 molar ratio in the feed stream on the mole fractions of tungsten carbides under the following conditions: reaction temperature of 1400°C, WCl_6 feeding rate of 0.6 g/min, $CoCl_2/WCl_6$ ratio of 1, C/W molar ratio of 2.7, and total flow rate of 1.1 L/min (25°C, 86.1 kPa). [Adapted from Ryu et al. (2008d)].

Effect of $CoCl_2$ concentration: The effect of $CoCl_2$ concentration was investigated by varying its feeding rate from 0 to 0.03 g/min, while maintaining all other conditions the same [reaction temperature of 1400°C, total flow rate of 1.1 L/min (25°C, 86.1 kPa), no H_2, CH_4 feeding rate of 0.01 L/min (25°C, 86.1 kPa), WCl_6 feeding rate of 0.06 g/min]. Figure 6.9 shows a plot of the $WC_{0.5}/WC$ molar ratio against the $CoCl_2/WCl_6$ molar ratio in the stream, which indicates that the degree of carburization increased with the increase of $CoCl_2$ concentration. This result can be explained by the catalytic effect of Co, as observed by Davidson et al. [1979]. The grain size of the product WC was 18 ± 1 nm when obtained without $CoCl_2$ addition, and increased to about 29 ± 1 nm when $CoCl_2$ was added at a rate of 0.03 g/min.

Effect of residence time: The effect of residence time on the formation of WC was investigated by varying the total flow rate of the gas mixture at the same composition of 0.38 mol % WCl_6 - 0.40 mol % $CoCl_2$ - 0.90 mol % CH_4 – Ar. All other conditions were kept the same (reaction temperature of 1400°C, no H_2, and C/W molar ratio of 2.3). Figure 6.10 shows the XRD patterns of the products obtained with different total flow rate of gases. The main product was still WC and Co. However, the amount of W_2C increased somewhat and even the formation of W and $Co_3W_9C_4$ occurred with increasing total flow rates, indicating that the degree of carburization decreased as the residence time decreased. The grain size of WC decreased from 25 ± 1 nm to 20 ± 1 nm as the total flow rate of gases increased from 1.1 L/min (25°C, 86.1 kPa) to 1.65 L/min (25°C, 86.1 kPa).

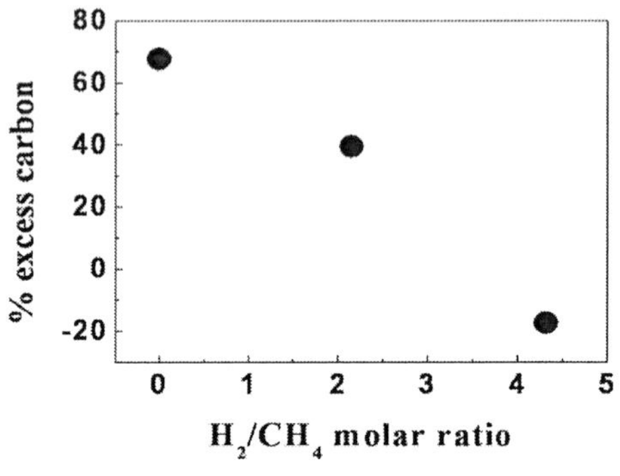

Figure 6.8. Effect of H_2/CH_4 molar ratio in the feed stream on % excess carbon. [Adapted from Ryu et al. (2008d)].

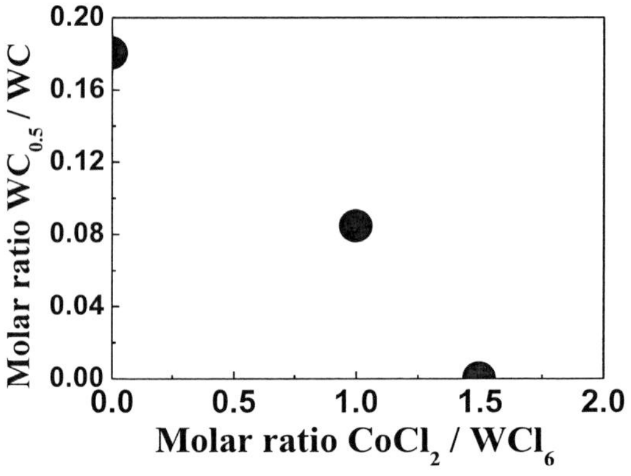

Figure 6.9. Effect of $CoCl_2$ concentration on the $WC_{0.5}/WC$ molar ratio under the following conditions: reaction temperature = 1400°C, total flow rate = 1.1 L/min (25°C, 86.1 kPa), CH_4 feeding rate = 0.01 L/min (25°C, 86.1 kPa), WCl_6 feeding rate = 0.06 g/min, and no H_2. [Adapted from Ryu et al. (2008c)].

Effect of reactants concentration: The effect of reactants concentration was investigated by varying the feeding rate of WCl_6 from 0.06 g/min to 0.6 g/min and that of $CoCl_2$ from 0.02 g/min to 0.2 g/min. Other experimental conditions were as follows: The WCl_6 volatilizer temperature was 440°C, the feeding rate of CH_4 was controlled to keep the C/W ratio in the feed at 2.3, the $CoCl_2/WCl_6$ ratio in the feed was 1, the reaction temperature was 1400°C, the total gas flow rate was made up by Ar to a constant value at 1.1 L/min (25°C, 86.1 kPa), and no H_2 was used in the feed stream since CH_4 alone gave high degree of carburization of W and was enough to reduce both chlorides [Ryu et al., 2008c]. Figure 6.11 shows the XRD patterns of the products obtained with different feeding rates of precursors. The product under these conditions was WC and Co together with small amounts of W_2C and/or W depending on the feeding rate of each reactant. No intermediate phases such as Co_3W_3C or Co_6W_6C were present in the product. Again, the feeding arrangement discussed in the previous section prevented the unreacted precursor gases from coming into contact with each other and forming intermediate phases, even when the concentration of precursors was increased.

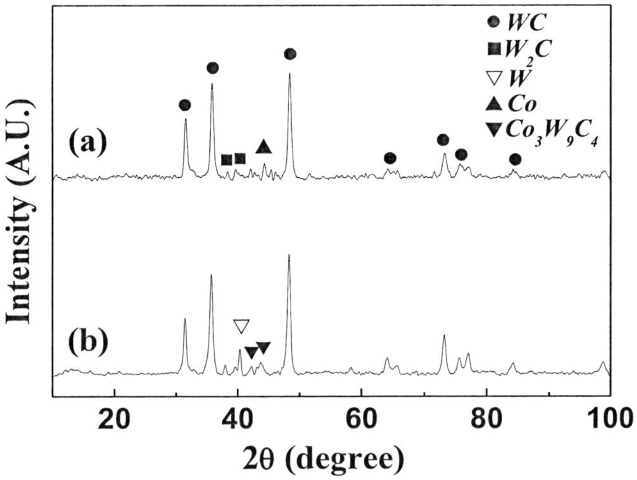

Figure 6.10. X-ray diffraction patterns of the product obtained at total flow rate of gases of (a) 1.1 L/min (25°C, 86.1 kPa) and (b) 1.65 L/min (25°C, 86.1 kPa), under the following conditions: reaction temperature = 1400°C, no H_2, and C/W (CH_4/WCl_6) molar ratio = 2.3. [Adapted from Ryu et al. (2008c)].

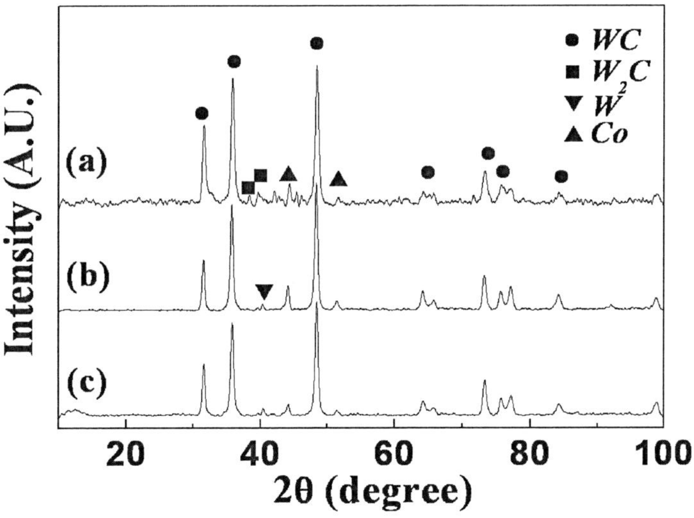

Figure 6.11. X-ray diffraction patterns of the products obtained with different feeding rates of precursors in the feed stream at (a) 0.06 g/min for WCl_6 and 0.02 g/min for $CoCl_2$, (b) 0.3 g/min and 0.1 g/min, and (c) 0.6 g/min and 0.2 g/min, respectively, under the following conditions: reaction temperature of 1400°C, WCl_6 volatilizer temperature of 440°C, C/W ratio of 2.3 in the feed stream, no H_2, and total flow rate of 1.1 L/min (25°C, 86.1 kPa). [Adapted from Ryu et al. (2008d)].

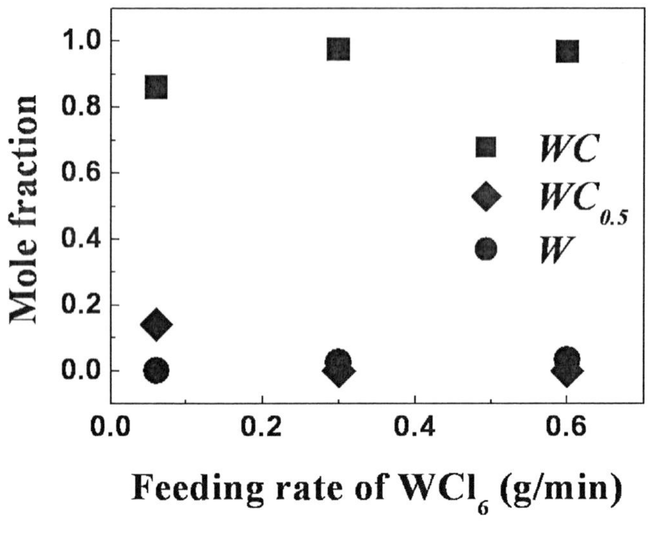

Figure 6.13. X-ray diffraction patterns of the products obtained with different WCl_6 volatilizer temperatures: (a) 440°C and (b) 540°C. [Adapted from Ryu et al. (2008d)].

Figure 6.12 shows the variation of the solid phase composition with that of the precursor feeding rate. The molar ratio was calculated from the XRD intensity of the WC peak at 2θ of 35.6, that of the W_2C peak at 2θ of 39.5, and that of the W peak at 2θ of 40.2, according to Equation (2). As the feeding rate of WCl_6 increased from 0.06 to 0.3 g/min, W_2C disappeared and a small amount of W was produced. The amount of W increased somewhat as the feeding rate increased to 0.6 g/min. The grain size of WC, which was 25 ± 1 nm at the lowest feeding rate, increased to 30 ± 1 nm at the highest feeding rate. The degree of carburization slightly decreased with an increase in reactant concentration. It is likely that the increase in WCl_6 concentration leads to increased grain size of the reduced W, which resulted in a slower carburization rate.

The C/W ratio was increased to 2.7 from 2.3 to enhance carburization degree since W_2C and/or W was observed at the latter ratio, even though the amount was small. The product contained negligible amounts of W_2C and/or W at this slightly increased C/W ratio, as seen in Figure 6.13 (a).

Effect of WCl_6 volatilizer temperature: The effect of WCl_6 volatilizer temperature was investigated by varying it from 440°C to 540°C while keeping all other conditions the same [no H_2, C/W ratio of 2.7, reaction temperature of 1400°C, total gas flow rate of 1.1 L/min (25°C, 86.1 kPa), WCl_6 feeding rate of 0.6 g/min, and $CoCl_2$ feeding rate of 0.2 g/min]. Figure 6.13 shows the XRD patterns of the products obtained with different volatilization temperatures of WCl_6. When WCl_6 volatilizer temperature was 440°C, the product contained negligible amounts of W_2C or W. However, when WCl_6 volatilizer was lowered into the reactor where the temperature was 540°C, the product contained small amounts of W_2C and W.

The grain size of produced WC was 30 ± 1 nm, which was not affected by the volatilization temperature of WCl_6 within the range tested. Each run was repeated 2 to 3 times, and the product grain size and the appearance of the small amounts of W_2C and W at the higher WCl_6 volatilization temperature were reproducible. No further work to definitively determine the reasons for this was carried out, however, because the difference was very small.

Post-Treatment of Produced Composite Powders

The effect of hydrogen heat treatment on the product composition, grain growth, and carbon content was investigated. The produced composite powder was placed in a ceramic boat, which was in turn placed in a tube furnace under hydrogen. The experimental conditions were as follows: The amount of powder treated was 2 g, the treatment temperature was 900°C, which was low enough to prevent rapid grain growth of particles, the heating rate of the tube furnace to 900°C was 10°C/min and the cooling rate was also 10°C/min, and hydrogen was flowed at a rate of 0.5 L/min (25°C, 86.1 kPa) after the temperature reached 900°C. The result showed that the unreacted W_2C and W phases were fully carburized to the WC phase after the treatment, as shown in Figure 6.14. The carburization of W_2C and W phases is likely caused by the highly mobile hydrocarbons, mainly CH_4, formed during the hydrogen heat treatment

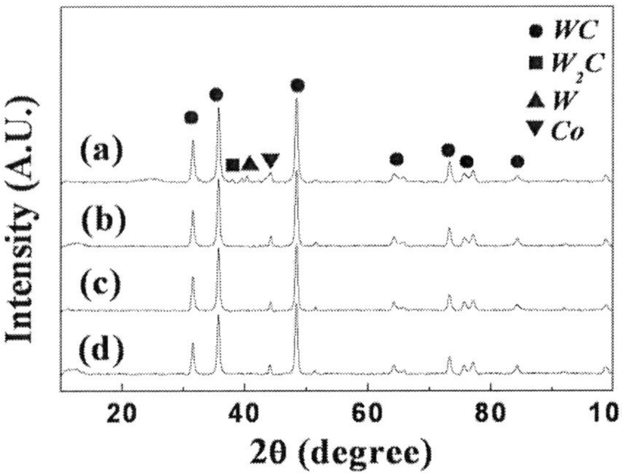

Figure 6.14. X-ray diffraction patterns of the products obtained with different hydrogen treatment times at 900°C using the powder produced from the tubular reactor: (a) before treatment, (b) 1 hour, (c) 2 hours, and (d) 4 hours of holding time. [Adapted from Ryu et al. (2008d)].

The % excess carbon in the product, 112 % before the treatment, was down to 7 % after 1 hour of post-treatment, 3 % after 2 hours, and 0 % after 4 hours, as shown in Figure 6.15. Up to 4 % excess free carbon is not only tolerated but often needed in the subsequent consolidation process to remove the small amount of oxygen present on the particle surface. Therefore, a post-treatment of 2 hours under the conditions used in this work is sufficient. The grain size of WC, 30 ± 1 nm before the treatment, only increased to 36 ± 1 nm after 4 hours of treatment. The particle size of composite powder before and after the hydrogen heat treatment was examined by TEM micrographs, as shown in Figures 6-16(a) and (b). The EDS maps of the composite powder obtained after the hydrogen heat treatment are shown in Figures 6-16(c) and (d). The particle size, as opposed to the grain size, of the composite powder was less than 40 nm before the treatment and became less than 70 nm after the treatment for 4 hours. It is noted that these values are from measurements on a rather limited number of particles. The EDS maps also indicate that WC and Co particles were uniformly mixed in the product, which verifies the significant feature of the chemical vapor synthesis (CVS) process to produce very uniformly mixed powders, as mentioned earlier.

To summarized, uniformly mixed WC-Co composite nanopowder was prepared by CVS in the WCl_6-$CoCl_2$-CH_4-H_2 system. The degree of carburization increased with increasing carbon potential and $CoCl_2$ concentration in the feed gas stream, in addition to increasing reaction temperature as discussed in Chapter 5.

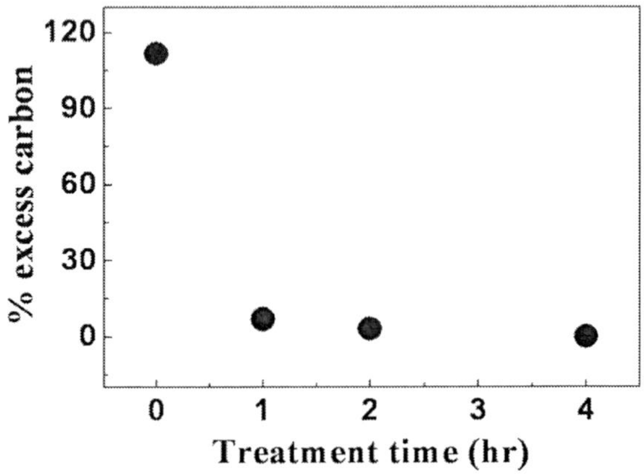

Figure 6.15. Effect of hydrogen heat treatment time at 900°C on % excess carbon. [Adapted from Ryu et al. (2008d)].

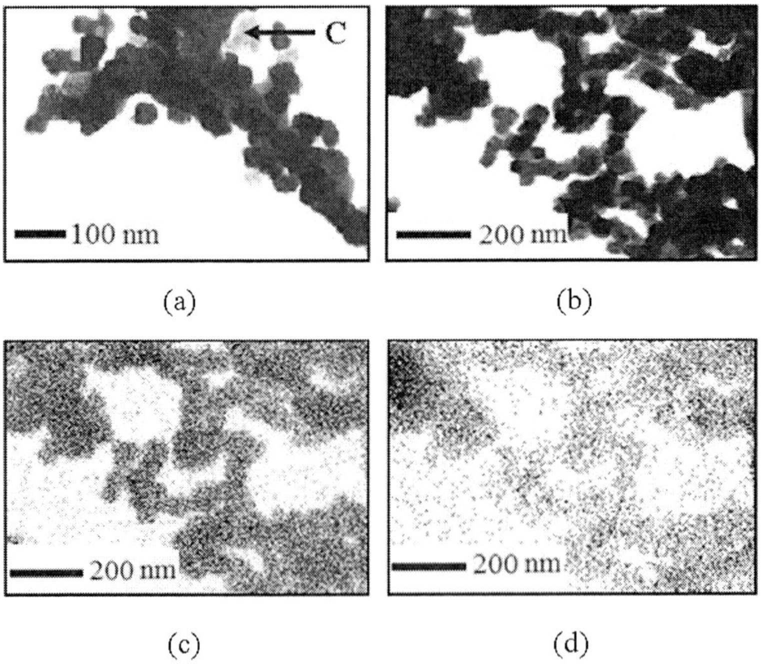

Figure 6.16. TEM photographs with EDS maps of WC-Co composite powders: (a) WC-Co composite powder obtained from the tubular reactor, (b) WC-Co composite powder after hydrogen heat treatment for 4 hours at 900°C, (c) EDS map of post-treated sample (Tungsten, black area), and (d) EDS map of post-treated sample (Cobalt, black area). [Adapted from Ryu et al. (2008d)].

The grain size of WC increased as $CoCl_2$ concentration and the residence time of WC increased, in addition to increasing reaction temperature as discussed in Chapter 5, but was not affected by carbon potential and H_2 addition within the ranges tested. The best conditions for synthesizing WC-Co composite powder in the apparatus used in this work were as follows: reaction temperature of 1400°C, CH_4 feeding rate of 0.01 L/min (25°C, 86.1 kPa), total flow rate of 1.1 L/min (25°C, 86.1 kPa), WCl_6 feeding rate of 0.06 g/min, and $CoCl_2$ feeding rate of 0.02 g/min.

Free carbon was always present in the product, because excess CH_4 was used. The amount of excess carbon in the composite powder was reduced as the hydrogen concentration was increased in the feed stream. By hydrogen heat treatment of the composite powder, the excess carbon in the product was completely removed and small amounts of unreacted W_2C and W phases were fully carburized to the WC phase. The cobalt content in WC-Co composite powder can also be adjusted by controlling the feeding rate of WCl_6 or $CoCl_2$ powder in the feed stream. In this work, the Co/W ratio in the composite product was 0.8, which is equivalent to the cobalt content of 19 wt % without free carbon after the post-treatment. Finally, WC-Co composite powder of particle sizes less than 70 nm was obtained. The CVS process carried out in our tubular reactor system confirmed that it is feasible to produce nanosized composite powders by this method.

6.2. Plasma-Assisted Synthesis

From APT and Co_3O_4

Ryu and coworkers [2009c] synthesized nanosized tungsten carbide-cobalt composite powder by a thermal plasma process, in which ammonium paratungstate and cobalt oxide were reacted with a gas mixture containing CH_4, H_2, and Ar. The reduction and carburization of vaporized precursors produced nanosized composite powder od tungsten carbide (WC_{1-x}) and cobalt, which contained small amounts of W_2C and/or W phase. The particle size of synthesized tungsten carbide powder was less than 20 nm. The synthesized composite powders were also subjected to a hydrogen heat treatment to fully carburize WC_{1-x}, W_2C, and W phases to the WC phase as well as to remove excess carbon in the product. Finally, nanosized WC-Co composite powder of particle size less than 100 nm was obtained.

An equilibrium composition diagram for the WO_3-Co_3O_4-H_2-CH_4 mixture was obtained by the use of HSC Chemistry software developed by Outotec [2002], as shown in Figure 6.17. The diagram shows that it is thermodynamically feasible to synthesize a mixture of WC and Co from the WO_3-Co_3O_4-H_2-CH_4 mixture in the plasma process.

Preliminary experiments were carried out to evaluate whether ammonium paratungstate (APT) and cobalt oxide (Co_3O_4) can be reduced in the plasma process before attempting to produce tungsten carbide-cobalt composite powder. Ammonium paratungstate [$(NH_4)_{10}W_{12}O_{41}\cdot xH_2O$, 99.97%] and cobalt oxide (Co_3O_4, 99%) of 40 μm average particle sizes were used as the precursors. Hydrogen (99.9%) and methane (99.9%) were used as the reducing and carburizing agents, respectively. Experimental conditions were as follows: Plasma torch power was 9 kW, the flow rate of plasma gas (Ar) was 28 L/min, feeding rate of pre-mixed APT-Co_3O_4 with a Co/W ratio of 0.7 was 3.5 g/min, and the flow rate of H_2 to carry APT-Co_3O_4 mixture was 4 L/min (25°C, 86.1kPa).

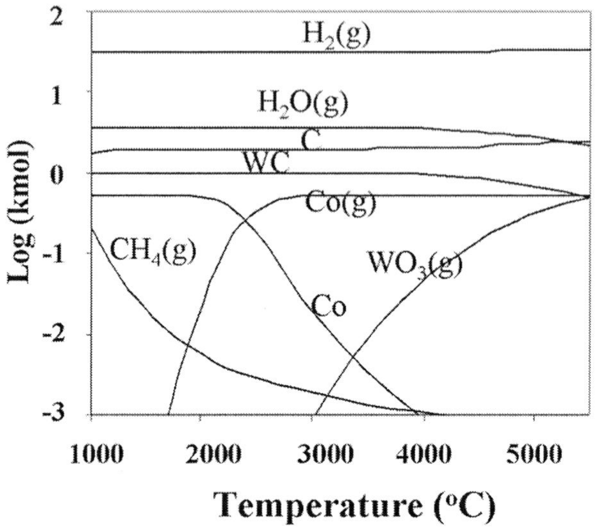

Figure 6.17. Calculated equilibrium composition of product from a mixture of WO_3, Co_3O_4, H_2, and CH_4 in the amounts of 1, 0.2, 30, and 3 kmol, respectively. [Adapted from Ryu et al., 2009c].

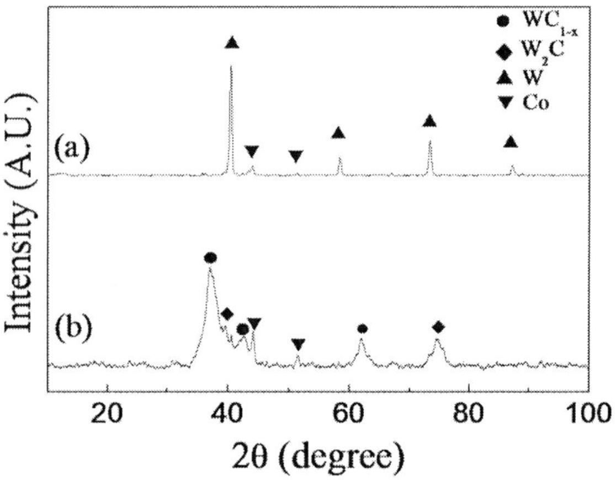

Figure 6.18. X-ray diffraction patterns of the products obtained by thermal plasma process from (a) APT-Co_3O_4-H_2 mixture and (b) APT-Co_3O_4-CH_4 mixture under the following conditions: Plasma torch power of 9 kW, the flow rate of plasma gas (Ar) of 28 L/min, feeding rate of pre-mixed APT-Co_3O_4 with a Co/W ratio at 0.7 of 3.5 g/min, and the flow rate of H_2 or CH_4 to carry APT-Co_3O_4 mixture of 4 L/min (25°C, 86.1kPa). [Adapted from Ryu et al., 2009c].

The product obtained under these conditions was W-Co composite powder, as shown in Figure 6.18(a). In subsequent experiments, CH_4 was fed at a flow rate of 4 L/min (25°C, 86.1kPa) with pre-mixed APT and Co_3O_4 powder to synthesize tungsten carbide-cobalt composite powder. The product was a mixture of WC_{1-x} and Co, which contained small amounts of W_2C and W phases, as shown in Figure 6.18(b). The effect of gas composition was also tested by varying the CH_4/H_2 ratio in the feed stream. However, the addition of H_2 in the feed stream resulted in the formation of substantial amounts of W_2C and W phases in the product. When CH_4 alone was used as a reaction gas, the highest degree of carburization was observed.

The vaporized APT and cobalt oxide were directly reduced and carburized at the high temperature of the plasma flame. No intermediate phases such as Co_3W_3C or Co_6W_6C were present in the product. The reason behind the formation of WC_{1-x} phase was explained in Section 5.2.

The grain size of WC_{1-x} calculated from XRD patterns by applying the Scherrer equation described in Section 3.2 was 9 ± 1 nm. EDS mapping of the composite powder was conducted and it confirmed that the WC_{1-x} and Co particles were uniformly mixed in the product, as shown in Figure 6.19. This presents another example of the significant feature of the vapor phase reaction to produce very uniformly mixed powders.

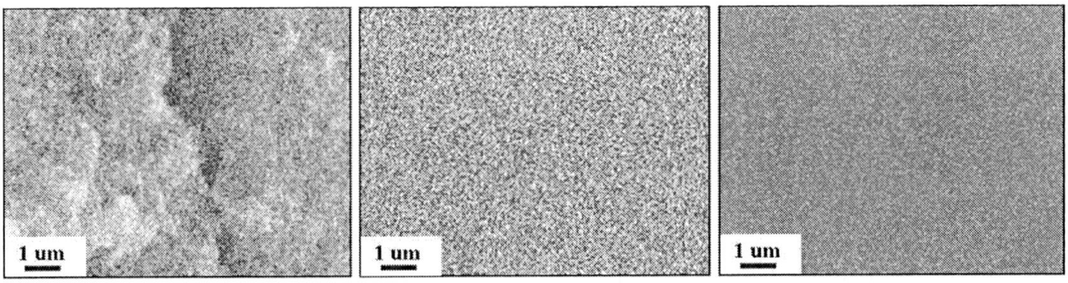

(a) (b) (c)

Figure 6.19. SEM image of WC_{1-x}-Co composite powder with associated EDS maps for tungsten and cobalt: (a) WC_{1-x}-Co composite powder, (b) EDS map of W (light area), and (c) EDS map of Co (gray area). [Adapted from Ryu et al., 2009c].

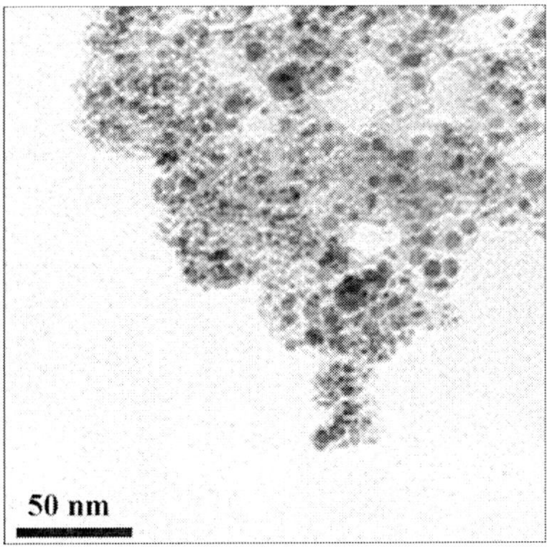

Figure 6.20. TEM micrograph of WC_{1-x}-Co composite powder synthesized under the following conditions: Plasma torch power of 9 kW, the flow rate of plasma gas (Ar) of 28 L/min, feeding rate of pre-mixed APT-Co_3O_4 of 3.5 g/min, and the flow rate of CH_4 to carry APT-Co_3O_4 mixture of 4 L/min (25°C, 86.1kPa). [Adapted from Ryu et al., 2009c].

The particle size of the synthesized composite powder was also examined using TEM micrographs. Figure 6.20 shows that the particle size ranged from 5 nm and 16 nm with an average size of about 10 nm. The particle size obtained from the TEM micrograph was close to the grain size calculated from the XRD results. Thus, the particles produced by the plasma process are single crystals. The shape of the produced particles was mostly round.

Post-Treatment of Produced Composite Powders

Free carbon was always present in the composite powder since excess methane was used in the reaction. By controlling the CH_4/H_2 ratio in the feed stream the amount of free carbon could be reduced. However, it also resulted in the formation of substantial amounts of W_2C and W phases.

The composite powders obtained from the plasma reactor were placed in a ceramic boat of 3.6 cm W x 9 cm L x 1.5 cm H, which in turn was placed in a tube furnace under H_2 atmosphere. The effects of hydrogen heat treatment on the product composition, grain growth, and carbon content were investigated. The amount of powder treated was 5 g, the treatment temperature was 900°C, the heating and cooling rates were 10°C/min, and hydrogen was flowed at a rate of 0.5 L/min (25°C, 86.1 kPa) after the temperature reached 900°C. This temperature was low enough to prevent rapid grain growth of particles but high enough to complete post-treatment reactions within a reasonable length of time. The unreacted WC_{1-x} and W_2C phases were converted to the fully carburized WC phase after the powder was treated by hydrogen for 2 hours, as shown in Figure 6.21. The % excess carbon in the product, 138 %, before the treatment, decreased to 37 % after 1 hour of post-treatment, and 0 % after 2 hours. The grain size of WC in the composite powder obtained after the treatment for 2 hours was 40 ± 1 nm. The particle sizes of composite powders after the treatment was also examined by TEM micrograph, which indicated that the largest particle size after the treatment for 1 hour was about 70 nm but most particles were smaller than 50 nm. When the composite powder was treated for 2 hours, the particle size increased but mostly remained at less than 100 nm, as shown in Figure 6.22.

In summary, nanosized tungsten carbide-cobalt composite powder was synthesized in a thermal plasma process, in which ammonium paratungstate (APT) and cobalt oxide were used as the precursors. The particle size of the composite powder, which mainly consisted of WC_{1-x} and Co particles, ranged from 5 nm to 16 nm with an average of about 10 nm. The synthesized composite powder was further subjected to a hydrogen heat treatment to fully carburize the WC_{1-x}, W_2C, and W phases to the WC phase and to remove excess carbon, which resulted in the formation of WC-Co composite powder. The particle size of the resulting composite powder was less than 100 nm. The thermal plasma process performed in this work confirmed its potential as an effective technique for the preparation of nanosized composite powders, owing to the high temperature to rapidly volatilize precursors followed by vapor phase reaction, and rapid quenching of the product to directly yield nanosized particles.

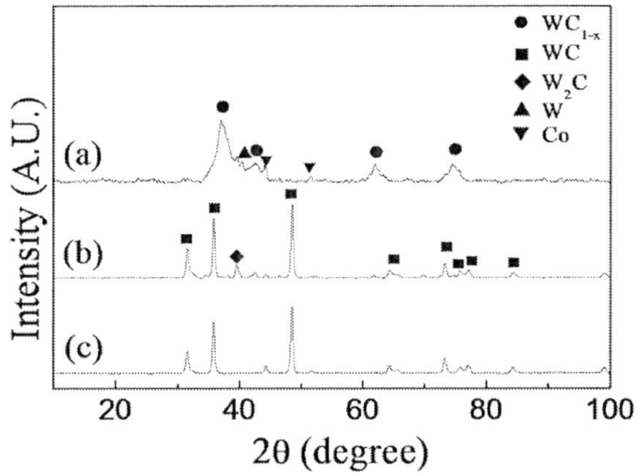

Figure 6.21. X-ray diffraction patterns of the products obtained with different hydrogen treatment times at 900°C using composite powder synthesized from a plasma reactor (a) before treatment, (b) 1 hour and (c) 2 hours. [Adapted from Ryu et al., 2009c].

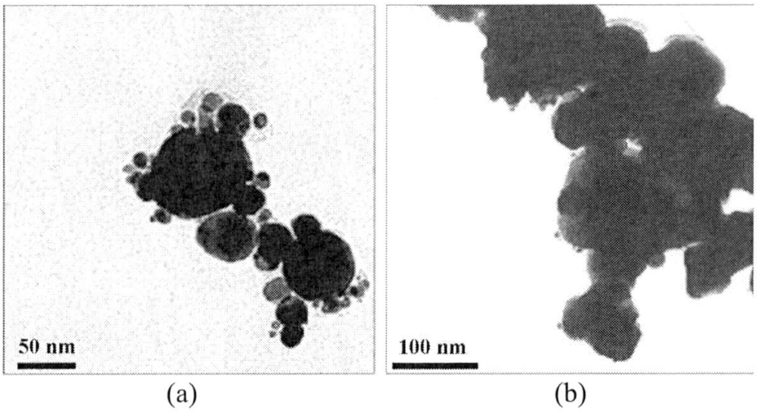

Figure 6.22. TEM micrographs of composite powder after (a) 1 hour and (b) 2 hours. [Adapted from Ryu et al., 2009c].

Synthesis of W-Co Followed by Separate Carburization

As seen in the previous section, the direct plasma synthesis of tungsten carbide-cobalt composite powder produces WC_{1-x}-Co mixtures that always contain excess carbon. The production of the WC_{1-x} phase, rather than the fully carburized WC, and the presence of excess carbon make post-treatment necessary. Ryu et al. [2010a] developed an alternate process of first producing uniformly mixed nanosized tungsten and cobalt composite powder followed by the subsequent carburization with CH_4. This resulted in the formation of nanosized WC-Co composite powder with a particle size less than 110 nm. This new method represents an alternate route that avoids the carbon removal step that is required when methane is used for direct carburization in the powder production step.

The details of the synthesis and properties of W-Co composite powder have been described in Section 3.5. This powder was placed in a ceramic boat, which in turn was placed in a tube furnace, and carburized by CH_4. The carburization of W may be achieved using carburizing agents such as carbon black, hydrocarbon gases, and carbon monoxide. Among these, carbon black powder is the most commonly used, by which it is easy to control the carbon content in the product. However, it requires a higher temperature (~1400°C), which results in a significant increase of grain size. When gaseous carburizing agents are used, the carburization temperature is lower, because of the efficient contact of gaseous molecules with tungsten particles. Methane is the commonly used gaseous carburizing agent because it is inexpensive and stable up to a high temperature [Gao and Kear, 1995].

The effect of carburization time on the product composition, grain growth, and carbon content was investigated at 900°C, which was low enough to prevent rapid grain growth of particles. Figure 6.23 shows the XRD patterns of the products obtained after the carburization at 900°C using powders synthesized in the plasma reactor. It is observed that tungsten was fully carburized to tungsten carbide after 1 hour.

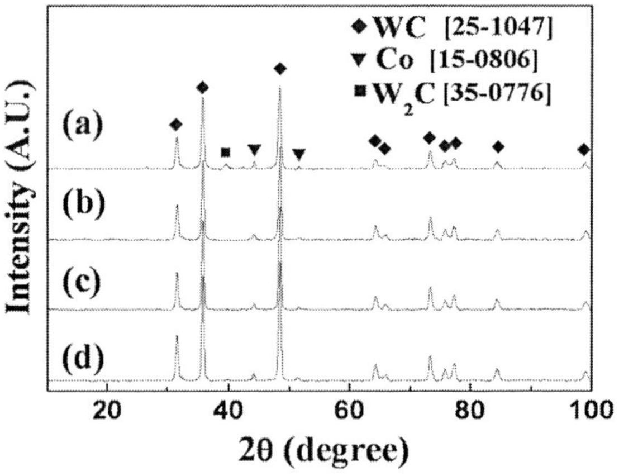

Figure 6.23. X-ray diffraction patterns of the products obtained after the carburization time at 900°C (a) 30 minutes, (b) 1 hour, (c) 2 hours, and (d) 3 hours. [The numbers after the phases are the JCPDS file numbers.] [Adapted from Ryu et al. (2010a)].

The grain size of produced WC was calculated from the XRD data by applying the Scherrer equation. Theoretically, the phase transformation from W to WC accompanies around 30 % lattice volume expansion, i.e. 20 nm W grains would become 22 nm WC grains. The experimental results indicate that the grain size of WC obtained was 38 ± 1 nm after 30 minutes of carburization time and it increased to 40 ± 1 nm and 46 ± 1 nm as the carburization time increased to 1 hour and 3 hours, respectively. Although there was some grain growth during carburization, the extent was rather moderate. The carbon content in the product was 96% of the stoichiometric amount for WC after a carburization time of 30 minutes and increased to 100% as the carburization time was increased to 1 hour. The fully carburized WC-Co composite powder had a Co content of 10 wt %. In the previous work [Ryu et al., 2009b], the post-treatment of the WC_{1-x} phase as the major phase obtained from the plasma reactor was conducted at 900°C under hydrogen atmosphere. To fully carburize it to the WC phase as well as to remove free carbon in the product, a treatment time of 3 hours was needed. The grain size of thus obtained WC phase was 43 ± 1 nm, which indicates an extensive grain growth considering that the grain size of the initial WC_{1-x} phase was 11 ± 1 nm.

The carburization temperature was lowered to 850°C because grain growth occurred during the treatment at 900°C. Figure 6.24 shows the XRD patterns of the products obtained from 1, 2, and 3 hours of carburization times at 850°C. The main product after the carburization at this temperature was WC and Co with a small amount of η-phase (Co_3W_3C). The amount of Co_3W_3C decreased as the carburization time was increased. The grain size of WC was 35 ± 1 nm after the carburization time of 1 hour at 850°C and increased to 40 ± 1 nm as the carburization time was increased to 3 hours. The carbon content in the product obtained after the carburization for 3 hours was 93%, indicating that W was still not fully carburized. Each of the runs was repeated 2 or 3 times and the results of the average grain size, carbon content, and product composition were reproducible. It can thus be concluded that a temperature of about 900°C is needed to fully carburize the plasma synthesized W powder within a reasonable amount of time and without excessive grain growths.

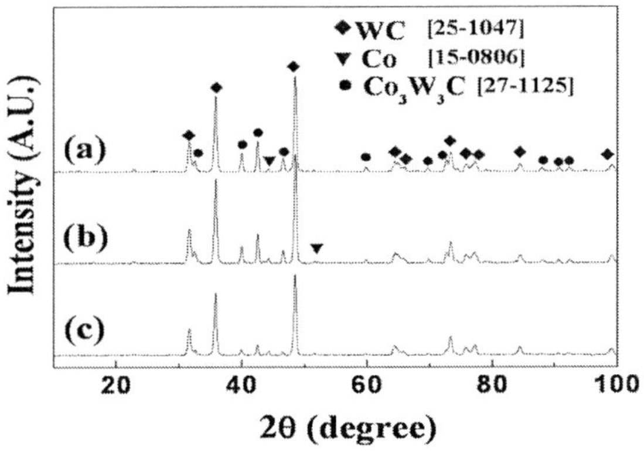

Figure 6.24. X-ray diffraction patterns of the products obtained after the carburization time at 850°C (a) 1 hour, (b) 2 hours, and (c) 3 hours. [The numbers after the phases are the JCPDS file numbers.] [Adapted from Ryu et al. (2010a)].

Figure 6.25 presents TEM micrographs of the particles obtained after carburization, which yield particle sizes in contrast to the grain sizes that XRD measurement produces. The average particle size of product obtained at the carburization temperature of 850°C for 3 hours was 52 nm ranging from 25 nm to 100 nm, based on the measurement from Figure 6.25(a). The particle size of product increased as the carburization temperature was increased to 900°C, as shown in Figures 6-25(b) and (c). The average particle size of product obtained after the carburization for 1 hour at 900°C was 60 nm and the sizes ranged from 35 nm to 110 nm. When the carburization time at this temperature was increased to 3 hours, the average particle size increased to 150 nm with a distribution from 100 nm to 230 nm. When these particle sizes are contrasted with the corresponding grain sizes, a moderate degree of agglomeration and chain formation at this temperature are indicated. Their extents were rather moderate, however, considering the very small particle size.

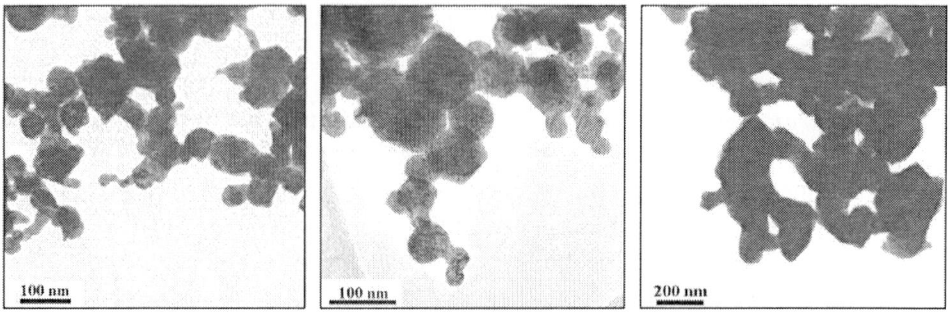

(a) (b) (c)

Figure 6.25. TEM micrograph of WC-Co composite powder synthesized under the following conditions: (a) 850°C for 3 hours, (b) 900°C for 1 hour, and (c) 900°C for 3 hours. [Adapted from Ryu et al. (2010a)].

To summarize, nanosized WC-Co composite powder was produced by a two-step processing via a thermal plasma process using ammonium paratungstate and cobalt oxide as the precursors. The W-Co nanocomposite powder obtained by the thermal plasma process showed very uniform distribution of W and Co particles with a Co/W molar ratio of about 0.4. The carburization temperature should be about 900°C to fully carburize W to WC by a mixture of CH_4 and H_2 within a reasonable amount of time and without excessive grain growth. Carburization at 900°C for 1 hour resulted in fully carburized WC-10 wt % Co composite powder of an average particle size of 60 nm with a size spread from 35 nm to 110 nm without excess free carbon. The WC particles are uniformly mixed with cobalt ready to be sintered rapidly into fully cemented tungsten carbide pieces.

Chapter 7

FLAME CHEMICAL VAPOR SYNTHESIS OF SILICA NANOPOWDER

ABSTRACT

Flame synthesis has been widely used in industry for the production of silica particles. Silica nanoparticles are commercially produced by the flame synthesis method because it can provide high production rates at relative low costs. The principle of this process, its applications and an example of experimental investigation are discussed in this chapter.

The flame process has been extensively used for the production of many ceramic and composite nanopowders [Pratsinis, 1998]. This process provides a good control of particle size and crystalline structure and can also continuously produce high-purity, thermally stable particles. It is an effective technique for producing fine, pure and single-phase particles in the as-prepared state [Jang, 1999; Mädler et al., 2002]. The sizes of the flame-synthesized particles range from a few to several hundred nanometers, depending on the operating conditions. Silica (SiO_2) nanopowder is one of the most commonly produced commodities of the flame process and has typically been obtained from the oxidation of silicon tetrachloride in a flame [Pratsinis, 1998]. Silica (SiO_2) nanoparticles have been widely used in various industrial applications, such as catalysis, pigments, pharmaceutics, electronic and thin film substrates, electronic and thermal insulators, and humidity sensors [Caruso et al., 1998; Giesche, 1994]. SiO_2 nanopowder is also nanometric fillers with potentially interesting reinforcing capabilities [Preghenella et al., 2005]. In these applications, the particle morphology, average particle size or specific surface area, and surface characteristics are the key characteristics of powders that must be controlled [Jang, 2001]. Silica powder has been synthesized by various methods including the sol-gel technique [Lu et al., 1997], surfactant-mediated processes [Inagaki et al., 2002], hydrothermal reactions [Tanev and Pinnavaia, 1996], chemical vapor deposition [Lee et al., 2005], and flame synthesis [Jang, 1999; Madler et al., 2002].

The flame synthesis method is useful for commercial production of SiO_2 nanopowder because of high production rates, relative low costs and the possibility for a continuous process. Jang and coworkers [Jang et al., 2006; Jang et al., 2007] have developed a flame process in which organic silicon precursors such as tetraethylorthosilicate (TEOS) and tetramethylorthosilicate (TMOS) obtained from the recycling of silicon sludge are used.

7.1. Synthesis of SiO_2 Nanoparticles from TEOS

Jang [2001] synthesized silica nanoparticles by the oxidation of tetraethyl-orthosilicate (TEOS) in a diffusion flame rector. This study reported on the results of using a bench-scale flame reactor with the specific purpose of increasing the typically low production rate of the flame process. A burner composed of five concentric tubes was developed for the diffusion flame, and used to produce several tens of grams of silica nanoparticles per hour. Silica nanoparticles ranging from 10 to 40 nm in average particle diameter were produced. The overall reaction for the vapor-phase synthesis of silica from TEOS is as follows:

$$Si(OC_2H_5)_4(g) + 12O_2(g) \rightarrow SiO_2(s) + 8CO_2(g) + 10H_2O(g) \tag{7-1}$$

In this work, a modified burner from a previous study [Jang, 1999] was used. The effects on the particle size of key variables such as TEOS concentration, total gas flow rate and gas composition in the flame were investigated. The amount of TEOS vapor was controlled by varying the temperature of the evaporation bath and the flow rate of argon. A diffusion flame burner (3.4 cm in outside diameter and 45 cm in length) composed of five concentric stainless tubes (10, 16, 22, 28, and 34 mm in outside diameter) that provided a stable flame over a wide range of operating conditions was used. A quartz tube of 10 cm in diameter and 120 cm in length was installed to guide the flame and particles. Hydrogen was used as the fuel while O_2 and air were used as the oxidants. Argon gas saturated with TEOS vapor from the evaporation bath was introduced to the central tube of the burner. The connecting line for TEOS from the evaporation bath to the burner was covered with a heating tape to prevent the condensation of TEOS vapor.

The average diameter from the TEM pictures was about 10 nm at the TEOS concentration of 9.80×10^{-6} mol/L and increased to 38 nm at 2.50×10^{-4} mol/L. The average diameter was also determined by BET analysis to be 11 nm at the TEOS concentration of 9.80×10^{-6} mol/L and 40 nm at 2.50×10^{-4} mol/L. As the total gas flow rate was changed from 40 to 20 L/min by decreasing the air flow rate through the 5th tube of the burner, the calculated adiabatic flame temperature of the flame increased from 2000 to 2500°C. Besides, the residence time of TEOS in the diffusion flame became longer. Such increases in the TEOS concentration, the flame temperature and the residence time are expected to result in an increase in particle size due to high coagulation, surface reaction, and the sintering rate of small particles. The average particle diameter of the silica nanoparticles from BET analysis indeed increased to 26 nm. The silica nanoparticles produced were amorphous. The effects of variations in the gas composition on the average particle diameter were rather complex, depending on the type of changes.

7.2. Synthesis of SiO_2 Nanoparticles Using TMOS Produced from Silicon Sludge

Tetramethylorthosilicate (TMOS) is one of the widely used organometallic compounds as precursors of silica based nanomaterials [Berggren et al., 2005; Kawakami et al., 2000; Dabadie et al., 1995]. Large amounts of waste silicon sludge that could serve as a useful resource for the production of SiO_2 nanopowder are generated from industry. The synthesis of silicon compounds such as TMOS from silicon powder by chemical reaction has seldom been reported [Mehrotra, 1988]. Jang et al. [2007] reported on the synthesis of aggregation-free silica nanoparticles by the flame spray pyrolysis (FSP) using TMOS synthesized from waste silicon sludge. The details of TMOS synthesis is given in the original reference. The effects of key variables such as the flow rates of hydrogen and dispersion air, the feed rate of precursor solution, and precursor concentration on the particle morphology and size were investigated. The morphology of as-prepared silica nanoparticles was spherical, non-hollow and non-aggregated. The average particle diameters ranged from 9 to 30 nm, being larger at a higher precursor feed rate and concentration.

As shown schematically in Figure 7.1, the apparatus for the synthesis of SiO_2 nanoparticles consisted of an ultrasonic atomizer for aerosol precursor, a diffusion flame burner, and a particle collector. The precursor solution was atomized with an ultrasonic vibrator (1.7 MHz), and the atomized droplets were subsequently carried into the central tube of the burner by a dry Ar gas flow. Hydrogen was used as the fuel while O_2 and air were used as the oxidants. The nanoparticles produced were collected with a bag filter.

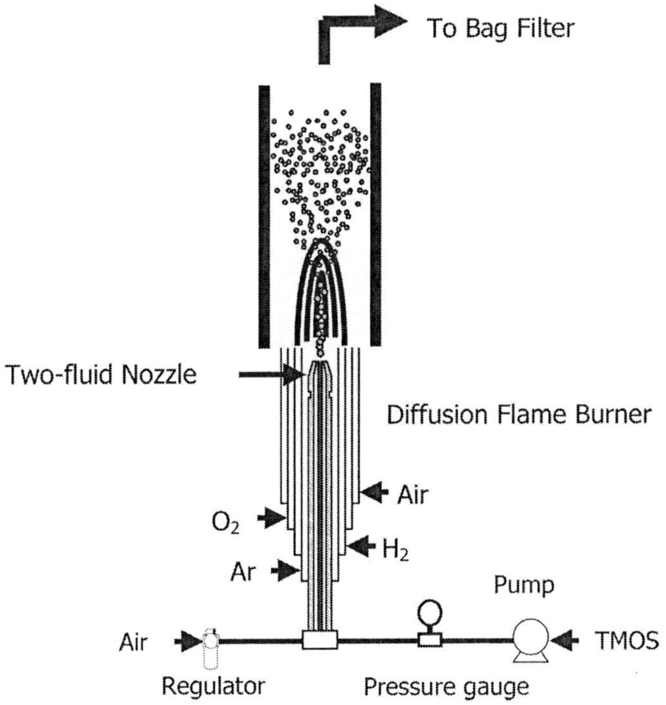

Figure 7.1. Schematic of experimental apparatus for the synthesis of SiO_2 nanoparticles. [Adapted from Jang et al. (2007)]

The particle morphology and size were characterized by transmission electron microscopy (TEM, Philips Model CM12), and the specific surface area of the particles was measured by nitrogen adsorption at -196°C using a BET method (Micrometrics Model ASAP 2400). An X-ray diffractometer (XRD, Rigaku Co. Model RTP 300 RC) was used to obtain the diffraction patterns of the synthesized particles.

$$Si(OCH_3)_4(g) + 6O_2(g) \rightarrow SiO_2(s) + 4CO_2(g) + 6H_2O(g) \tag{7-2}$$

The prepared SiO_2 nanoparticles were amorphous based on the analysis of X-ray diffraction pattern. Figure 7.2 shows TEM micrographs of the prepared SiO_2 nanoparticles at different concentrations of the precursor (TMOS) at otherwise fixed process conditions [TMOS aerosol (8.3 L/min of dispersion air + 26.2 mL/min of TMOS), argon (5 L/min), hydrogen (4 L/min), oxygen (10 L/min) and air (35 L/min), and air (60 L/min)]. The average particle diameters determined by BET analysis ranged from 9 nm to 13 nm as the TMOS concentration increased from 30 to 90% by volume.

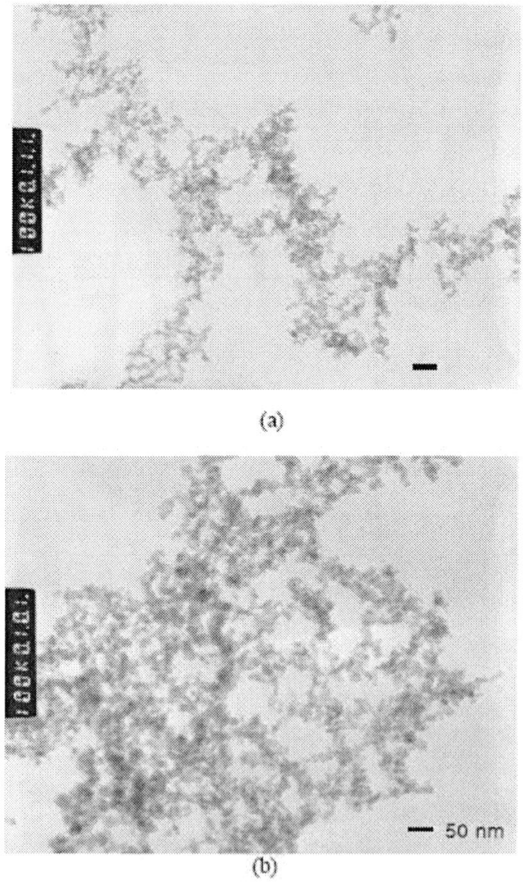

Figure 7.2. TEM images of SiO_2 nanoparticles synthesized from different TMOS concentrations by flame spray pyrolysis ((a): 30 vol%, (b): 90 vol%) [Adapted from Jang et al. (2007)].

The morphology of the prepared silica nanoparticles was spherical, non-hollow and there was no neck formation when the concentration of the precursor was higher than 50 %. However, when the concentration was as low as 30%, chain-like aggregates were generated. When the liquid droplets size is sufficiently small, the droplets are evaporated instantly due to the Kelvin effect [Friedlander, 2000]. The evaporated substance undergoes chemical reaction, nucleation, condensation, collision, and coalescence, finally producing the chain-like aggregates. Besides, the effect of precursor feed rate on the particle morphology and size was investigated by increasing it from 18.3 mL/min to 35.8 mL/min while keeping the flow rates of dispersion air and other gases constant. Even when the precursor feed rate was increased, the morphology of the prepared powder remained spherical, non-hollow and non-aggregated, but the average diameter of particles increased from 10 nm to 15 nm.

As the flow rate of dispersion air was decreased from 4.0 L/min to 8.3 L/min at otherwise fixed experimental conditions, the specific surface area of the silica nanoparticles decreased from 240 to 95 m^2/g and the corresponding average particle diameter increased from 11 nm to 30 nm.

To summarize, flame synthesis of silica nanoparticles from TMOS was conducted by the use of a reactor equipped with a two-fluid nozzle. The average particle size of the synthesized powder ranged from 10 to 60 nm and the particle morphology was spherical, non-aggregated, and non-hollow. The average particle size increased with precursor feed rate, hydrogen flow rate, and precursor concentration.

7.3. Flame Vapor Synthesis vs. Flame Spray Pyrolysis

Flame synthesis processes can be divided into the flame vapor synthesis (FVS) and the flame spray pyrolysis (FSP). In FVS, the precursor is introduced into the flame as a vapor. The important steps in FVS are chemical conversion of the precursor to oxide, the nucleation of oxide particles, the collisions of these particles, the condensation of oxide vapor onto the particles, and the sintering of the particles [Friedlander, 2000]. However, FVS has the disadvantages of requiring vapor-phase precursors. Also, it is not easy to control the concentration of each precursor during the synthesis of multi-component particles. In FSP, the precursors are fed into the flame reactor as sprayed droplets. Important advantages of FSP over FVS are that it can easily synthesize multi-component particles and use precursors with a wide range of volatility [Chang et al., 2006; Jang et al., 2004; Mueller et al., 2004; Sahm et al., 2004; Strobel et al., 2004; Madler and Pratsinis, 2002; Madler et al., 2002; Tani et al., 2002; Bickmore et al., 1998; Pratsinis, 1998; Bickmore et al., 1996). FSP is more complicated than FVS in terms of particle formation mechanism because the evaporation of the droplets must be considered in addition to the process steps in FVS.

Many different precursors including hexamethyldisiloxane (HMDSO) [Madler et al., 2002], tetraethylorthosilicate (TEOS) [Jang et al., 2006], tetramethylsilane (TMS), tetramethylorthosilicate (TMOS) and tetrapropylorthosilicate (TPOS) [Jossen et al., 2005] have been used to form silica particles by FSP. However, these precursors have relatively low boiling points ranging from 27 to 275°C. Hence, the morphology of silica particles made by FSP using them was similar to that made by FVS. Furthermore, there have been no systematic studies to determine the effect of precursor volatility on the product morphologies and the particle formation mechanisms. Morphology is an important factor in the application of particles. Aggregated particles are preferred in reinforcing materials such as carbon black for tires. On the other hand, spherical particles are desirable for pigments in paint and carrier particles in drug delivery.

Cho et al. [2009] synthesized silica particles by FSP from two precursors, TEOS and silicic acid ($[SiO_x(OH)_{4-2x}]_n$), which have significantly different volatilization characteristics. They further present two mechanisms, considering the different evaporation and particle formation times, to explain the different morphologies of the particles obtained from the two precursors. Tetraethylorthosilicate (TEOS, 98%, ACROS Organics, USA) was mixed with ethanol (DC Chemical Co., 99.5%) to several concentrations. Aqueous silicic acid solution was prepared from sodium silicate solution by an ion exchange process with cation exchange resin to remove sodium ions. A schematic diagram of the experimental apparatus for the preparation of SiO_2 powder is shown in Figure 7.3. The experimental apparatus consisted of an ultrasonic nebulizer (Alfresa Pharma Co., UN-511), a diffusion flame burner, and a particle collector. The precursor solution was sprayed by a vibrator of the ultrasonic nebulizer, operated with a frequency of 1.7 MHz. Droplets sprayed from the precursor solution were carried by Ar gas (2 L/min) into the central tube of the diffusion burner with an outer diameter of 36 mm and a length of 300 mm. The liquid flow rate was 1.07 mL/min which varied somewhat depending on the viscosity of the liquid. Additional Ar gas (1 L/min) was flowed through the second tube from the center of the burner for a stable and straight flow of the flame. H_2 (99.995%, 5 L/min) was used as the fuel while O_2 (99.95%, 6 L/min) and air (15 L/min) were used as the oxidants.

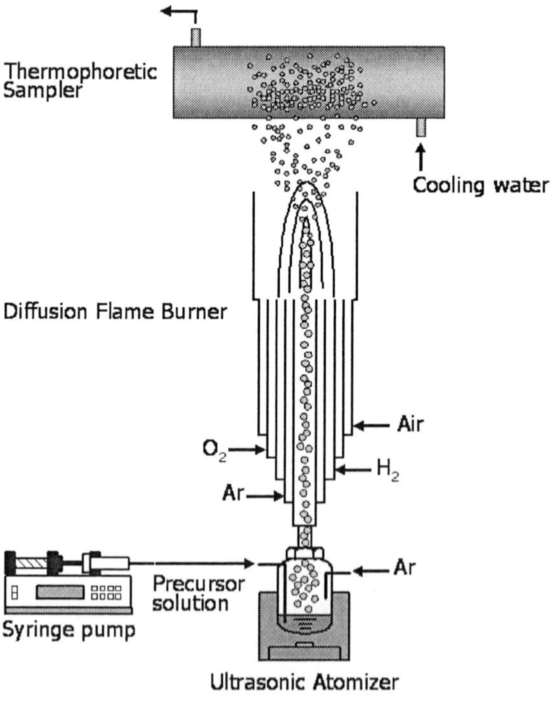

Figure 7.3. Schematic diagram of the experimental set-up. [Adapted from Cho et al. (2009)].

Figure 7.4 shows the temperature profile in the flame. It was measured by using a Pt-Rh R-type thermocouple with a 0.5 mm diameter. In all the runs, the flame temperature was measured three times along the axial and radial directions of the burner without feeding the precursor. The particles generated in the flame were collected by thermophoresis on the surface of a water-cooled glass tube (100 mm in diameter and 300 mm in length) installed 150 mm above the top of the burner.

The particle morphology was examined by transmission electron microscopy (TEM, Philips, CM 12). The specific surface area of the particles was measured by nitrogen adsorption at -196°C using a BET instrument (Micrometrics, Tristar 3000). The average particle diameter was calculated from the specific surface area of the particles using the relationship between the specific surface area (A) and the particle diameter (d_p), $d_p = 6/(\rho_p \cdot A)$, where ρ_p is the density of the particle (SiO_2, $\rho_p = 2.2$ g/cm^3), assuming spherical particles.

The crystallinity of the particles were analyzed by X-ray diffractometry (XRD, Rigaku, RTP 300 RC) with CuKα target operated at 30 kV and 40 mA. The chemical bonds of the particles were analyzed by a Fourier transform infrared spectroscopy (FT-IR, Thermo Electron, Nicolet 380). XRD and FR-IR data showed only silica peaks. Hence, the synthesized particles were deemed to be highly pure silica. The major difference between TEOS and silicic acid is volatility. The boiling point of TEOS is around 169°C, whereas there is no boiling point of silicic acid because it readily forms silica when it decomposes while losing water. Volatility can affect the evaporation rate of the precursors and the formation of particles in the flame. A precursor with a high evaporation rate in FSP forms particles similar to that in FVS. A precursor with a low evaporation rate might form particles different from that with a high evaporation rate.

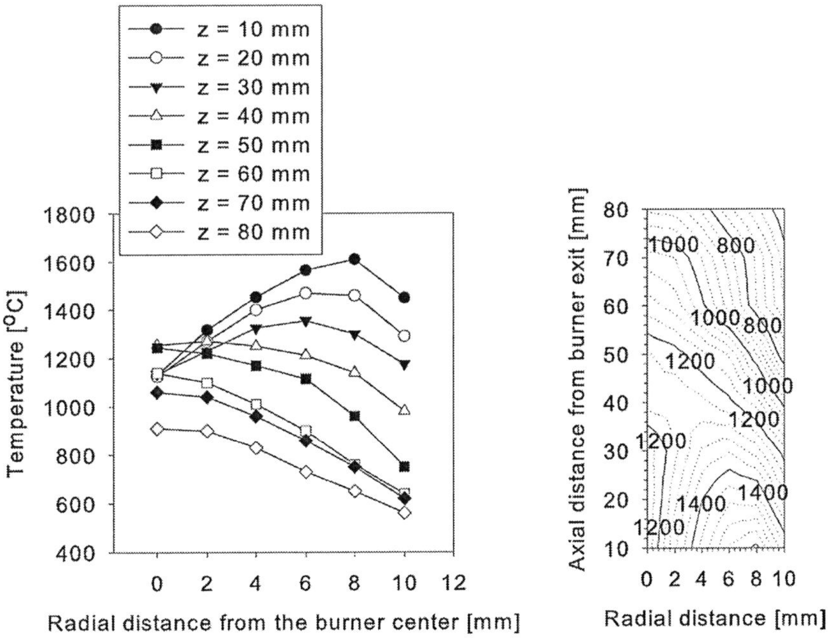

Figure 7.4. Temperature profile of the flame. [Adapted from Cho et al. (2009)].

Effect of Precursor Concentration

The concentration of the precursors was varied from 0.1 to 0.5 M to examine the effect of evaporation rate on the particle morphology. The maximum flame temperature was fixed at 1610°C. As the TEOS concentration in the precursor solution increased from 0.1 to 0.5 M, the specific surface area of silica powder decreased from 285 to 81.4 m^2/g and the average particle size calculated from it increased from 9.6 to 33.5 nm (Figure 7.5a). These results reveal that the average particle size can be controlled by changing the precursor concentration in FSP. However, particles did not grow in proportion to the increase in the precursor concentration. The average particle size increased 3.5 times when the precursor concentration increased 5 times. If one particle was generated from one precursor droplet, the average diameter of particles for a precursor concentration of 0.5 M would be $5^{1/3}$ = 1.71 times of the average diameter of particles from a precursor concentration of 0.1 M [Messing et al., 1993]. Hence, the factor of 3.5 might indicate there are two different particle formation mechanisms depending on the concentration of the precursor. To elucidate this phenomenon, the particle morphology was examined by TEM. Figure 3.5b-d shows TEM micrographs of the silica particles prepared by the FSP using TEOS solutions of various concentrations. The morphology of the prepared silica nanoparticles was spherical and solid. There were almost no aggregated particles when the concentration of TEOS was 0.5 M. On the contrary, only chain-like aggregates were produced when the precursor concentration was 0.1 M. Particles made by FVS generally form chain-like aggregates. When the precursor concentration was 0.3 M, both aggregated and dispersed particles coexisted. These results indicate that silica nanopowder from 0.1 and 0.5 M TEOS solutions was formed by different mechanisms. When the precursor concentration was 0.1 M, TEOS was completely evaporated in the flame. The evaporated precursor formed silica aggregates. When the precursor concentration was 0.5 M, TEOS droplets were 1.71 times larger than those generated from the 0.1 M solution. The evaporation time of droplets is proportional to the square of the droplet size [Hinds, 1999]. Hence, the evaporation time of the droplets formed from the 0.5 M TEOS solution is 2.92 times longer than that from the 0.1 M TEOS solution. The thermal decomposition of TEOS has previously been expressed as a first order reaction [Chu et al., 1991; Herzler et al., 1997]. Since the droplet temperature is the same if there is no superheating, the particle formation time is the same regardless of the precursor concentration. Therefore, the increased evaporation time of droplets may prevent TEOS from completely evaporating, resulting in different particle morphologies.

Wegner and Pratsinis [2003] obtained more and larger particles at higher production rate of 25 g/h or lower temperature in flame vapor synthesis. They reported that the particles were formed by the fusion of agglomerates. However, the droplet evaporation was not instantaneous when the precursor concentration was high in the experiments of Cho et al. [2009]. If coalescence was the only mechanism for large and spherical particles, there should be larger spherical particles at a longer residence time and higher temperature. Figure 7.5e shows the specific surface areas of silica sampled from various heights from the top of the burner at the TEOS concentration of 0.3 M. The result showed no correlation of particle size with residence time. Furthermore, the results obtained under various flame temperatures did not support coalescence for larger spherical particles. Based on the above analysis, they believed that particles were formed inside the droplets before the droplets evaporated completely in the case of the higher-concentration TEOS solution.

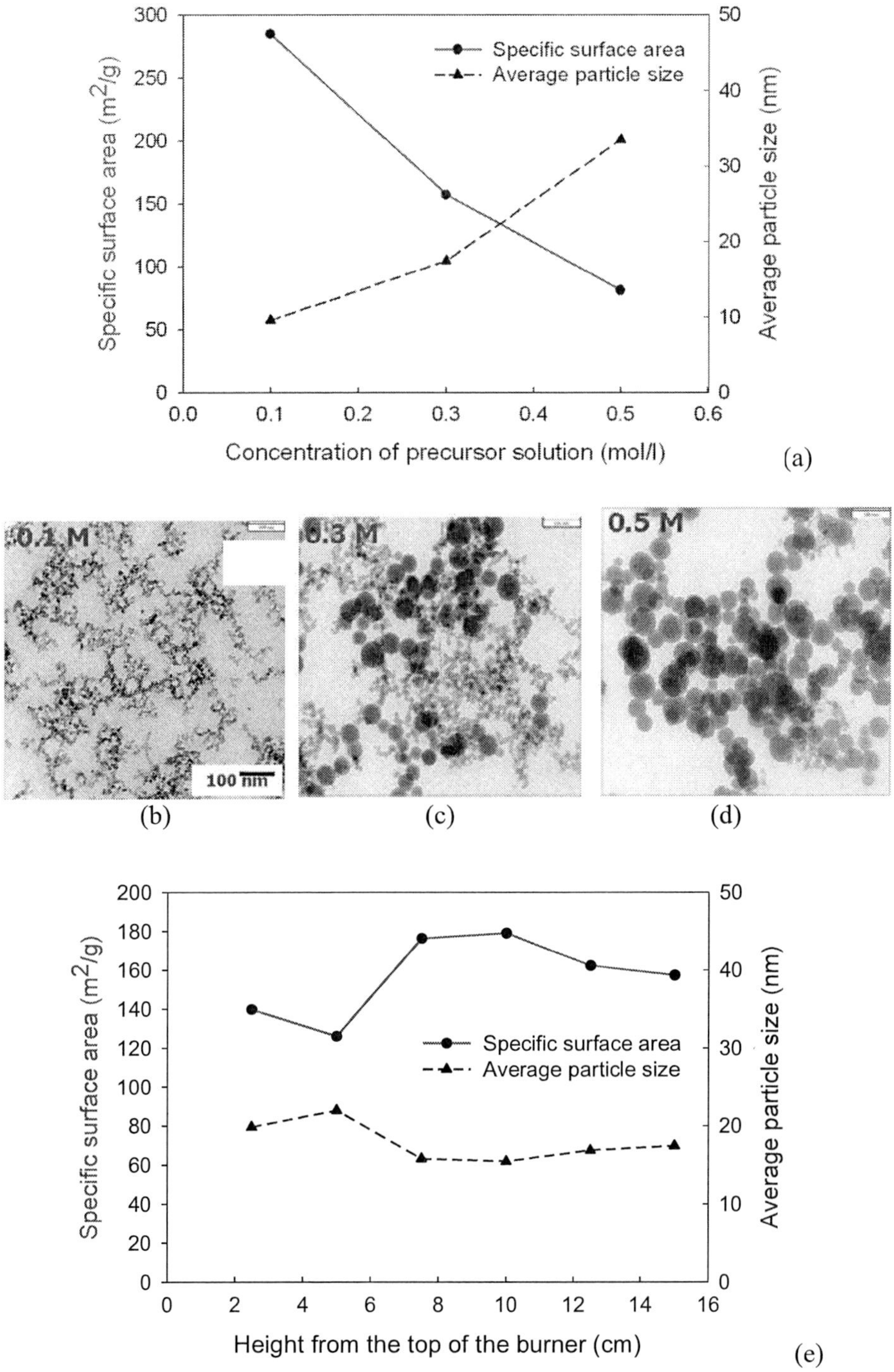

Figure 7.5. (a) Specific surface areas and average particle sizes of silica particles made from TEOS at various concentrations at 1610°C. TEM micrographs of silica particles for (b) 0.1, (c) 0.3 and (d) 0.5 M. (e) Specific surface areas and average particle sizes of silica particles produced from TEOS at various heights from the top of the burner at 0.3 M and 1610°C. [Adapted from Cho et al. (2009)].

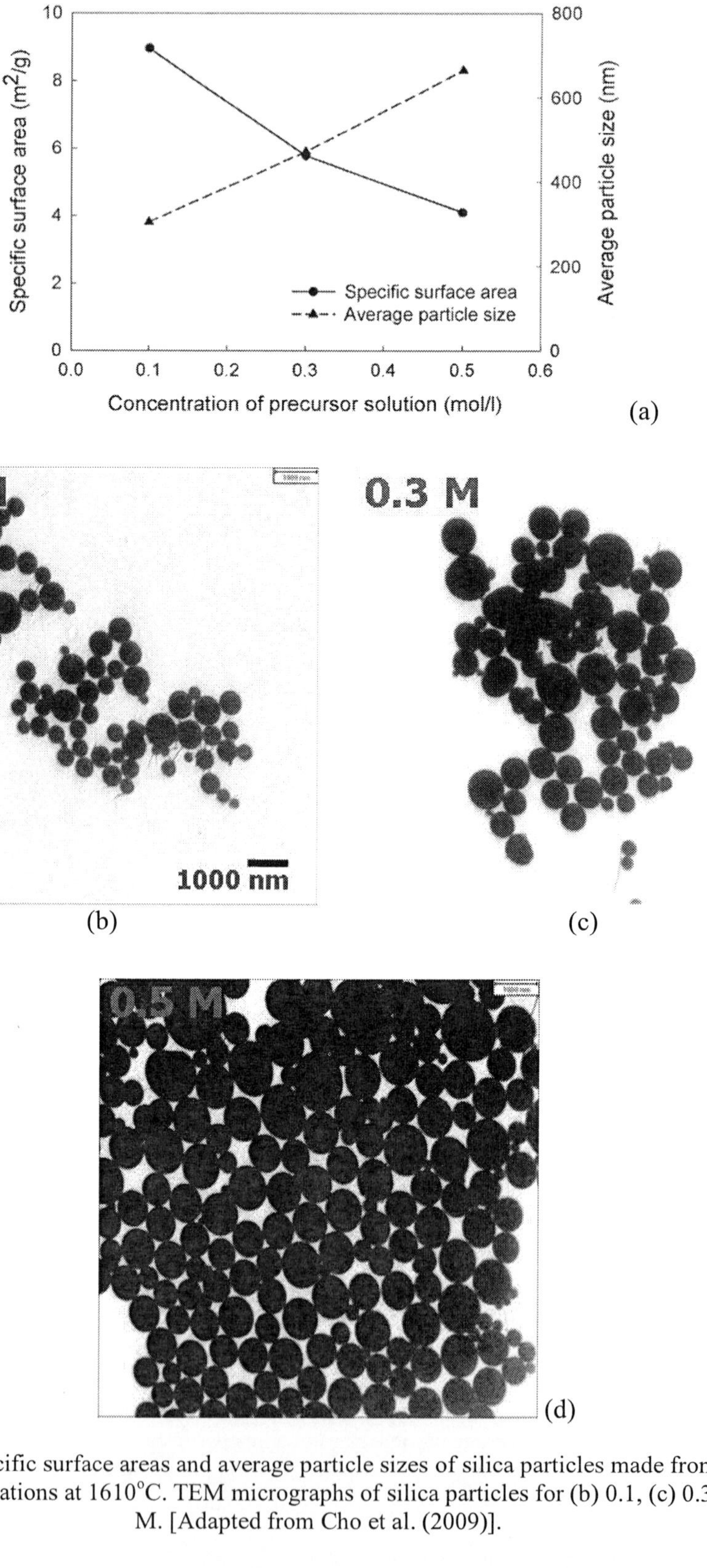

Figure 7.6. (a) Specific surface areas and average particle sizes of silica particles made from silicic acid at various concentrations at 1610°C. TEM micrographs of silica particles for (b) 0.1, (c) 0.3 and (d) 0.5 M. [Adapted from Cho et al. (2009)].

Silica particles were also prepared from aqueous silicic acid under the same conditions. The specific surface area values of the prepared powders from silicic acid solutions of 0.1, 0.3 and 0.5 M concentrations were 8.95, 5.78 and 4.10 m^2/g, respectively. The average particle diameter calculated from these values increased from 305 to 472 to 665 nm, respectively (Figure 7.6a). The actual increase of particle diameter between the first and the third cases was 2.18 times which was close to 1.71 compared to the case for TEOS. The difference in the increase factor may be due to the difference in droplet size. The droplet sizes were calculated from the particle size with the assumption of one particle formation from one droplet. They were 2.18, 2.34 and 2.78 μm for 0.1, 0.3 and 0.5 M, respectively. The increase of droplet size with the concentration might be due to the increase of viscosity [Lefebvre, 1989] and/or due to the polymerization of silicic acid [Iler, 1979]. The calculated sizes of the droplets were smaller than the measured [Wang et al., 2008b]. This was due to the disintegration of droplets in the flame. However, the sizes of disintegrated droplets were larger than those with acetate compounds in FSP [Jang et al., 2004]. Polymerization of silicic acid might lessen the disintegration of the droplets.

Figures 7-6 b-d shows TEM pictures of particles from silicic acid solutions of various concentrations. At all concentrations, spherical, non-hollow and isolated particles were formed. This implies that there was only one mechanism to form silica particles from silicic acid, unlike in the case of TEOS. As stated earlier, the evaporation time of silicic acid was very long due to its low volatility. Hence, silicic acid formed one particle from one droplet resulting in only isolated particles regardless of its concentration.

Effect of Flame Temperature

If the particles were made from the precursor vapor, an increase in the flame temperature would promote the particle growth due to higher coalescence rate resulting in larger primary particle size [Friedlander, 2000]. But Cho et al. [2009] investigated another effect of evaporation time of the precursor TEOS on the particle morphology with respect to flame temperature when there was incomplete droplet evaporation at a high precursor concentration. The maximum flame temperature in this work was varied from 1200 to 1710°C by controlling H_2 flow rate from 3 to 6 L/min while the precursor concentration in the solution was held at 0.5 M. The average size of the particles made from TEOS decreased from 47.6 to 15.9 nm as the maximum flame temperature increased from 1200 to 1710°C (Figure 7.7a). TEM pictures of silica particles prepared at different flame temperatures are shown in Figures 7-7 (b)-(e). Although the flame temperature increased, the morphology of as-prepared particles remained spherical and non-aggregated when the maximum flame temperature was lower than 1610°C. However, when the maximum flame temperature was 1710°C, both non-aggregated particles and chain-like aggregates were formed, as shown in Figure 7.7e. The evaporation rate is proportional to temperature [Friedlander, 2000]. Hence, the evaporation time decreased as the temperature increased. The particle formation time inside the droplets, however, was similar because temperature of the droplets remained the same around the boiling point of TEOS. Therefore, as temperature increased, evaporation became faster than particle formation. At the higher flame temperature, precursor vapors converted to particles after the evaporation of precursor. When the flame temperature was lower, particles formed inside the precursor droplet which did not evaporated completely. This is why the morphologies of particles were different at different temperatures when the precursor concentration was high.

The specific surface area and average size of particles from a silicic acid solution were not changed noticeably with respect to the maximum flame temperature (Figure 7.8a). In addition, the morphologies of as-prepared silica particles were spherical, non-hollow and non-aggregated under all experimental conditions (Figures 7-8 b-e). Because silicic acid does not evaporate, silica particles were formed directly from the precursor droplets and thus the specific surface area was independent of temperature. The silica particles from both the TEOS and silicic acid solutions were amorphous, regardless of temperature.

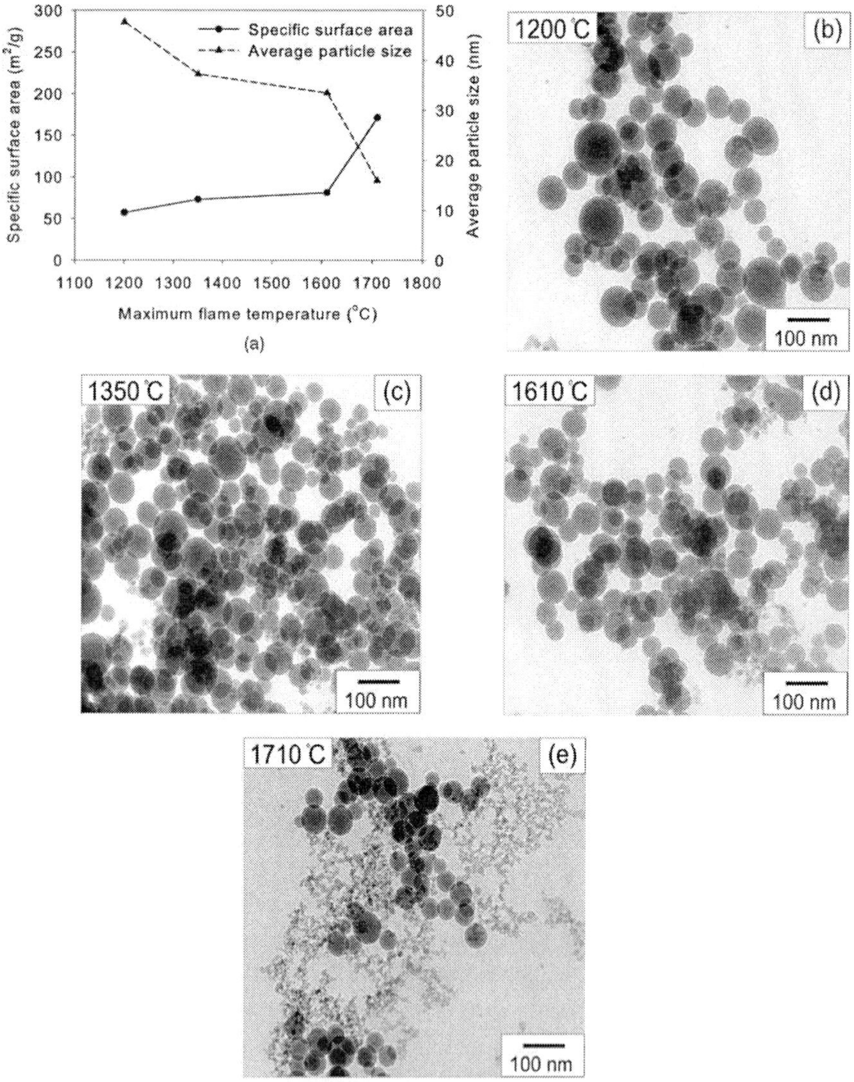

Figure 7.7. (a) Specific surface areas and average particle sizes of silica particles made from TEOS at various temperatures and the precursor concentration of 0.5 M. TEM micrographs of silica particles obtained at (b) 1200, (c) 1350, (d) 1610 and (e) 1710°C. [Adapted from Cho et al. (2009)].

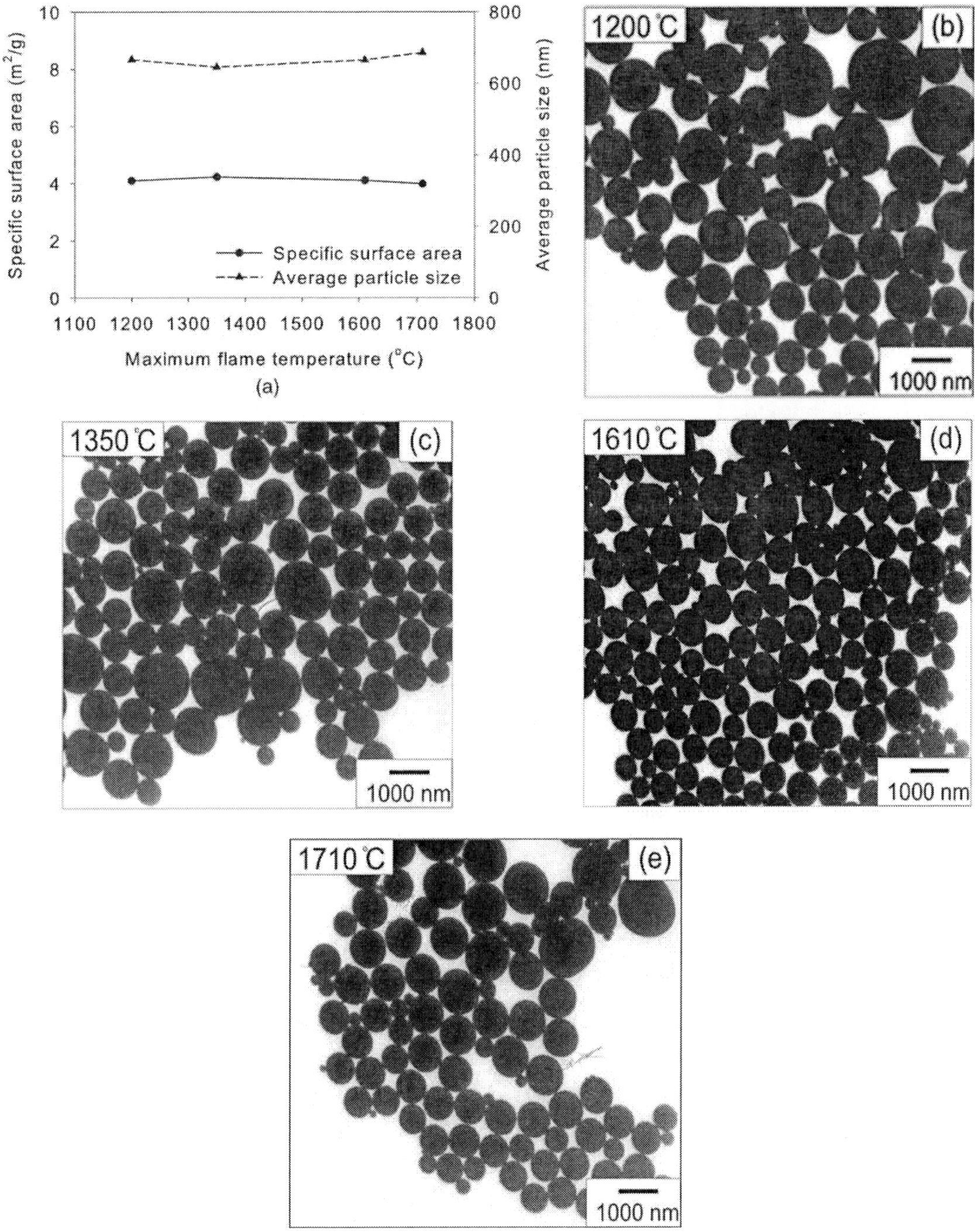

Figure 7.8. (a) Specific surface areas and average particle sizes of silica particles made from silicic acid at various temperatures and the precursor concentration of 0.5 M. TEM micrographs of silica particles obtained at (b) 1200, (c) 1350, (d) 1610 and (e) 1710°C. [Adapted from Cho et al. (2009)].

Particle Formation Mechanisms

The paths illustrated in Figure 7.9 are proposed as the mechanisms for the birth and growth of SiO_2 particles from sprayed droplets of precursors by FSP. Primary droplets are formed from the ultrasonic nebulizer. The average size of the droplets are around 6 μm [Wang et al. 2008b]. These primary droplets disintegrate in the flame due to a large temperature gradient. The size of disintegrated droplets was around 2 μm as calculated from the results for silicic acid. The solvent, such as water or ethanol, evaporates first after the disintegration, leaving the precursor in the droplet, because the solvents have lower boiling points than the precursor. Then, there are two main mechanisms for particle formation and growth based on the evaporation time and particle formation time. Definitive comparison of the times could not be made because data on particle formation kinetics, vapor pressure at the droplet surface, and the diffusion coefficient of precursor vapor are limited.

The kinetics of silica formation from precursor droplets is not fully understood, unlike from the vapor [Chu et al., 1991]. The kinetics of particle formation is not dependent on the flame temperature because the temperature of a burning droplet is fixed at its boiling point if there is no superheating. Hence, the particle formation time is almost fixed. However, the evaporation time is dependent on the precursor concentration and flame temperature.

The first mechanism is the vapor phase reaction [VPR, (A) in Figure 7.9] for the case when the evaporation time of the precursor is much shorter than the particle formation time by thermal decomposition or precipitation. In this case, primary particles are formed after the complete evaporation of the precursor droplet. The experimental conditions of low precursor concentration and high flame temperature (e.g., 0.1 M at 1610°C) of TEOS belong in this case. The primary particles formed from the precursor vapors typically grow into small and aggregated particles by nucleation, condensation and coagulation in the flame.

The second mechanism, involving intradroplet reaction [IDR, (C) in Figure 7.9], is for the opposite case when the evaporation time of the precursor is much longer than the particle formation time. It applies to all the cases for silicic acid. In this case, seed particles are created by precipitation inside the precursor droplet before its evaporation. They then form one particle in the droplet by condensation and coagulation. Finally, large and non-aggregated spherical particles are formed.

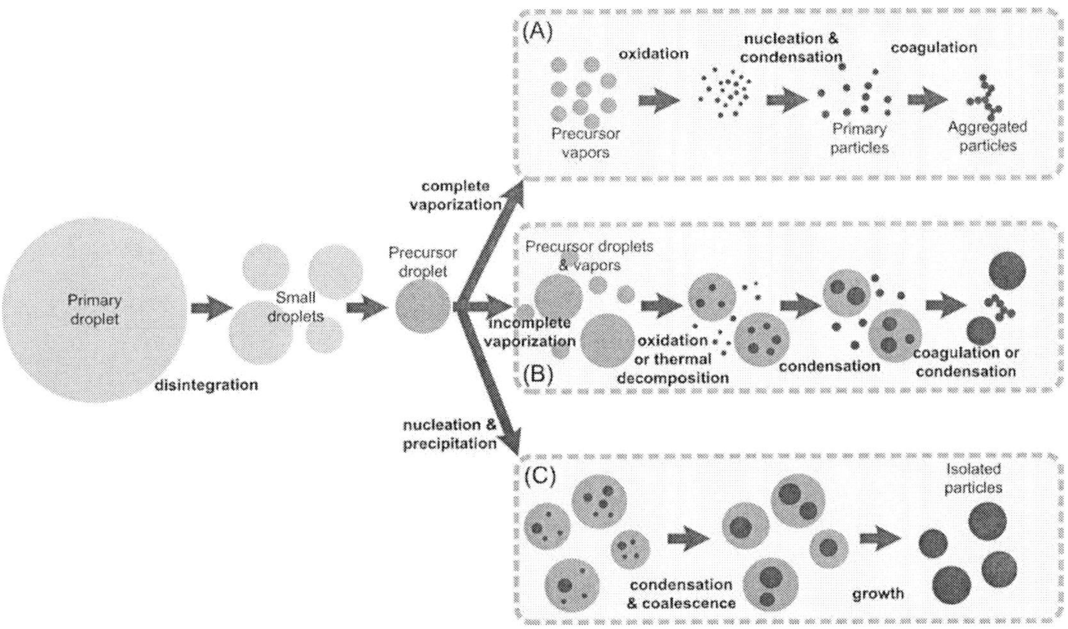

Figure 7.9. Particle formation mechanisms. (A) vapor phase reaction (VPR), (B) mixture of VPR and intra-droplet reaction (IDR), and (C) IDR. [Adapted from Cho et al. (2009)].

When the evaporation time and particle formation time are comparable, a combination of VPR and IDR occurs, as shown in Figure 7.9(B). It occurs for a higher precursor concentration (e.g., higher than 0.3 M at 1610°C) and a lower flame temperature in the case of TEOS (e.g., lower than 1710°C at 0.5 M). Particles form inside the droplets by thermal decomposition while the droplets are evaporating. Thus, two types of particles are produced: small and aggregated particles made by VPR and larger and spherical particles made by IDR. However, the sizes of the particles made from TEOS by IDR were smaller than that from silicic acid due to the partial evaporation of TEOS. After particles form by VPR and IDR, they may experience coagulation.

To summarize, silica particles having different morphologies were synthesized by FSP from two precursors, TEOS and silicic acid, with different volatilities. The morphology of the synthesized particles depended on the evaporation rate of the precursor. When the evaporation time was shorter than the particle formation time as in FVS, aggregated particles were formed. Spherical and isolated particles were formed under the opposite condition. Based on these observations, two types of particle formation mechanism, VPR and IDR, were proposed. By choosing an appropriate precursor, particle morphology as well as their crystallinity and size can readily be controlled by taking advantage of the knowledge of these different mechanisms.

Chapter 8

PLASMA-ASSISTED CHEMICAL VAPOR SYNTHESIS OF ZIRCONIA NANOPOWDER

ABSTRACT

The application of plasma-assisted chemical vapor synthesis (CVS) to the preparation of advanced oxide nanopowders is illustrated in this chapter with the synthesis of zirconia nanopowder as an example.

Zirconia (ZrO_2) is a stable oxide that is widely used in many industrial applications such as ceramics, oxygen sensors, fuel cells, and resistive heating, thanks to its superior properties of high melting point, low thermal conductivity, good toughness, high chemical stability, high oxygen-ion conductivity, and high refractive index [Chraska et al., 2000; Mayo et al., 1999; Gell, 1995; Cutler et al., 1992; Gleiter, 1992]. Nano-sized zirconia powders have been produced by various methods such as mechanical milling, sol-gel method, precipitation method, and vapor-phase process [Adam et al., 2008; Tartaj et al., 2007; Stenger et al., 2005; Mueller et al., 2004; Mädler et al., 2002; Bondioli et al., 2001; Kilian and Morse, 2001; Hu et al., 1999; Tani et al., 1983]. The vapor-phase process has the advantage of producing nano-sized powder with de-agglomerated particles and a narrow size distribution, and avoids the multiple steps, high liquid volumes, and chemical reagents required in the wet chemical processes.

Zirconia can be obtained by the decomposition of zirconium carbonate above 600°C [Gimblett et al., 1988]. However, it is not possible to produce nano-sized zirconia powder with conventional vaporization and condensation methods because of its high boiling temperature. Ryu et al. [2010b] synthesized nano-sized zirconia and yttria-stabilized zirconia powders by applying the thermal plasma process described in Section 3.2 and Figure 3.1. In their work zirconium carbonate was decomposed and subsequently vaporized in the high temperature plasma flame to produce nano-sized zirconia powder. The effects of the plasma torch power and plasma gas flow rate on the product phase was determined. Further, the effect of yttria doping on the thermal stability of the zirconia phase was investigated.

The plasma reactor system used by Ryu et al. [2010b] was equipped with a plasma generator, powder supply unit, cylindrical reactor, cooling system, precursor feeding system, and powder collector. The precursors used for synthesizing zirconia and yttria-doped zirconia were: (1) zirconium carbonate ($2ZrO_2 \cdot CO_2 \cdot xH_2O$) powder for the former, and (2) a mixture of zirconium carbonate ($2ZrO_2 \cdot CO_2 \cdot xH_2O$) and yttrium carbonate ($Y_2(CO_3)_3 \cdot xH_2O$) powders for the latter. Each precursor was milled in a jar mill to prepare particles in the size range of 3-10 μm. The milled powder was injected to the plasma flame through a feeding port of the plasma torch using the powder feeder described in Section 3.2 [Ryu et al., 2009a; 2008a]. The produced zirconia powder was collected using a Teflon-coated polyester filter with a pore size of 1 μm.

The powder was characterized by Transmission Electron Microscopy (TEM, JEOL, JEM-2000FXII) equipped with an Energy Dispersive X-ray Spectrometer (EDS) for elemental analysis, a BET surface area analyzer (Micrometric, ASAP 2010), an X-Ray Diffractometer (XRD, Siemens D 5000), a Raman Microscope Spectrometer (RMS, FT-Raman) for phase identification, and ZetaPALS (Brookhaven Instruments) to measure particle size. The grain size of the synthesized powders was calculated by the Scherrer equation from the XRD pattern [Cullity, 1978]. The weight fractions of monoclinic and tetragonal phases of ZrO_2 were estimated from the XRD patterns by using the following formula [Howard and Hill, 1991; Schmid, 1987]:

$$X_M = \frac{I(111)_M + I(11\bar{1})_M}{I(111)_M + I(11\bar{1})_M + I(111)_T} \tag{8-1}$$

where the subscripts M and T refer to the monoclinic and tetragonal phases, respectively, and I(hkl) refers to the integrated intensity of the corresponding peaks.

The effects of plasma torch power in the range of 7-34 kW on product phase and grain size were investigated under the following experimental conditions: (1) plasma gas (Ar) flow rate = 39 L/min (25°C, 86.1 kPa total pressure at Salt Lake City), (2) precursor feeding rate = 2 g/min, (3) flow rate of Ar to carry the precursor powder = 4 L/min (25°C, 86.1 kPa), and (4) pressure of reaction chamber = 86.1 kPa.

The XRD patterns of the products obtained at different levels of plasma torch power is shown in Figure 8.1. The product was a mixture of monoclinic and tetragonal phases of ZrO_2. The weight fraction of the monoclinic phase, calculated from Eq. (8-1), decreased from 0.64 to 0.26 with the plasma torch power in the range tested, as shown in Figure 8.2.

The zirconia phases formed at different temperatures can be predicted from the phase diagram [Scott, 1975]. The cubic, tetragonal and monoclinic phases form at 2750°C, 2370°C and 1170°C, respectively, as molten zirconia cools. The presence of the metastable tetragonal phase in this study is caused by the rapid quenching in the plasma reactor. Further, no cubic phase was observed in this study, even with the rapid quenching. This is likely due to the fast transition of cubic phase to tetragonal phase [Gibson and Irvine, 2001; Sheu et al., 1992].

The grain sizes of monoclinic and tetragonal phases of ZrO_2 were estimated from the Scherrer equation applied to the XRD peaks to be 17 ± 1 nm and 16 ± 1 nm at 7 kW, respectively, and 22 ± 1 nm and 21 ± 1 nm at 34 kW, indicating that the grain sizes increased somewhat with torch power. The particle size was also examined from the TEM micrographs. The average particle size of the powder obtained at 7 kW was 17 nm with a range from 10 nm to 37 nm, as shown in Figure 8.3(a), based on the manual measurement of a rather limited number of particles. When the plasma torch power was increased to 34 kW, the average particle size increased to 23 nm with a range from 15 nm to 38 nm, as shown in Figure 8.3(b).

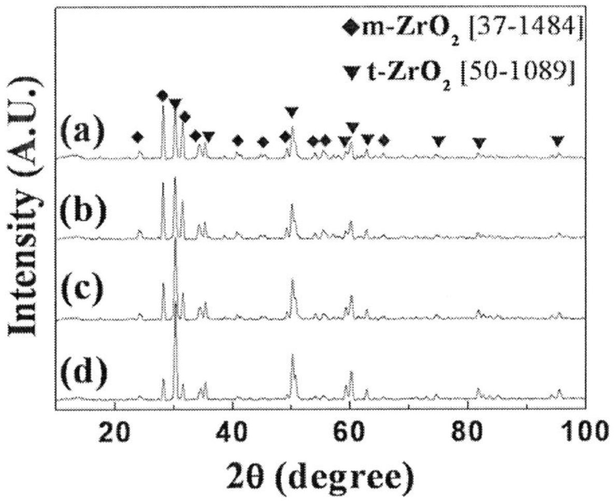

Figure 8.1. X-ray diffraction patterns of the products obtained with different power levels of plasma torch: (a) 7 kW, (b) 13 kW, (c) 28 kW, and (d) 34 kW. [The numbers after the phases are the JCPDS file numbers.] [Adapted from Ryu et al. (2010b)].

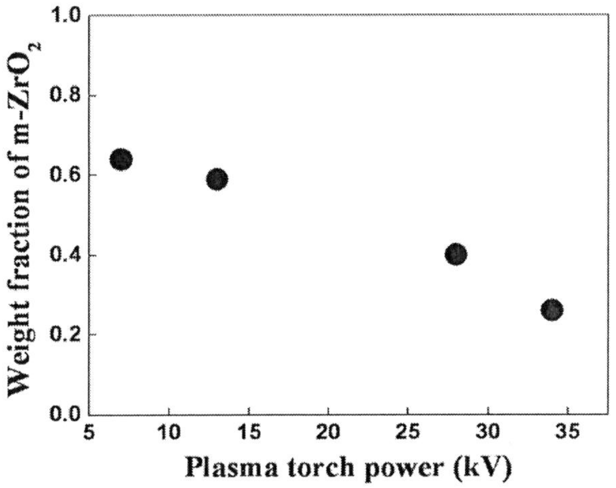

Figure 8.2. Weight fraction of monoclinic phase in the product obtained with different power levels of plasma torch. [Adapted from Ryu et al. (2010b)].

The effect of the plasma gas (Ar) flow rate on the product phase and grain size was determined by varying the rate from 29 L/min to 57 L/min (25°C, 86.1 kPa) under otherwise identical conditions [plasma torch power = 27 kW, precursor feeding rate = 2 g/min, and flow rate of Ar to carry the precursor powder = 4 L/min (25°C, 86.1 kPa)]. The product obtained under these conditions was a mixture of monoclinic and tetragonal phases of ZrO_2. The weight fraction of monoclinic zirconia phase at the gas flow rate of 29 L/min (25°C, 86.1 kPa) was 0.48 which decreased to 0.32 at the gas flow rate of 57 L/min (25°C, 86.1 kPa), as shown in Figure 8.4. The increased gas flow rate provided a shorter time for the transformation of the tetragonal phase to the monoclinic phase. The grain sizes of monoclinic and tetragonal phases of ZrO_2 powders estimated from the XRD patterns were 20 ± 1 nm and 19 ± 1 nm, respectively, indicating that the grain size was not affected by the plasma gas flow rate within the range tested.

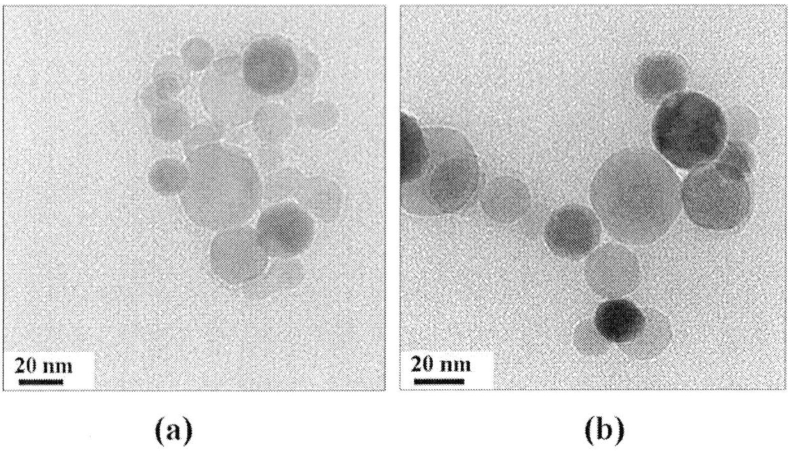

Figure 8.3. TEM micrographs of zirconia nanopowders synthesized with various power levels of plasma torch: (a) 7 kW and (b) 34 kW. [Adapted from Ryu et al. (2010b)].

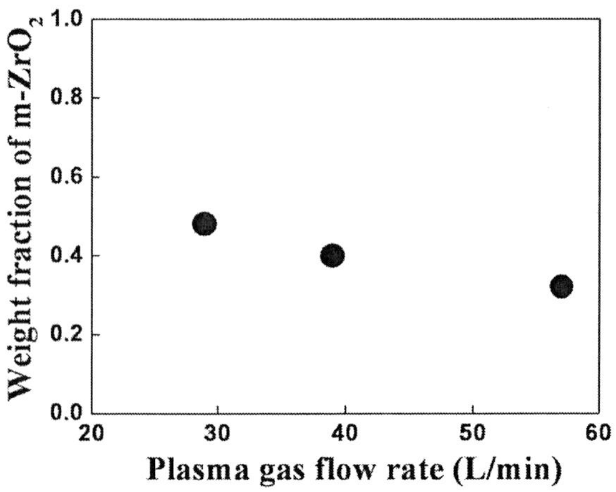

Figure 8.4. Weight fraction of monoclinic phase in the product obtained with different plasma gas flow rates. [Adapted from Ryu et al. (2010b)].

As discussed above, the production of tetragonal phase, an important material in practical applications, was increased by controlling the torch power and/or the flow rate. However, pure tetragonal phase was not obtained under the test conditions. It is known that metastable tetragonal phase tends to transform into monoclinic phase when heated [Kiminami, 1990]. Thus, Ryu et al. [2010b] focused on the formation of stabilized tetragonal phase, as discussed in following section.

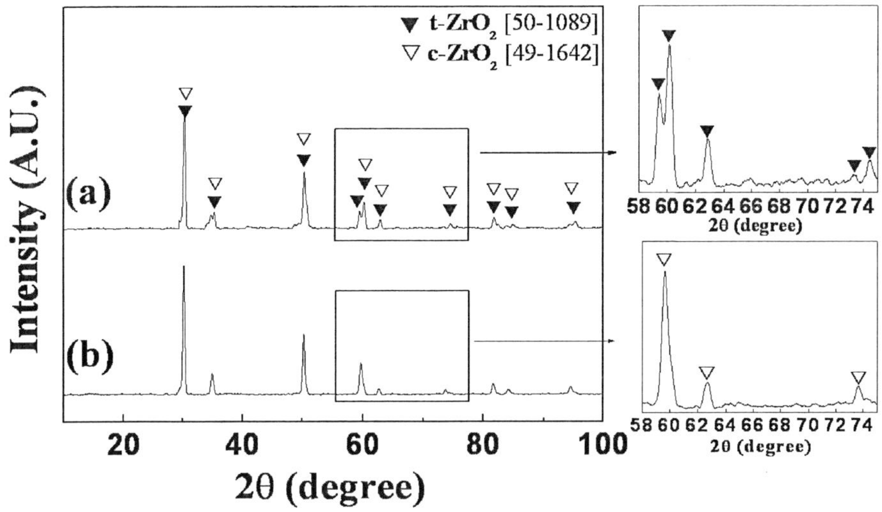

Figure 8.5. X-ray diffraction patterns of the products with different yttria contents: (a) 4YSZ and (b) 8YSZ. [Adapted from Ryu et al. (2010b)].

8.1. Synthesis of Yttria-Stabilized Zirconia Nanopowder

The tetragonal and cubic phases of ZrO_2 can be stabilized by the addition of yttria [Vasylkiv et al., 2001; Tsukada et al., 1999; Ramamoorthy et al., 1995; Ruh et al., 1984; Scott, 1975]. It was reported [Scott, 1975] that the tetragonal phase with less than 8 mol% yttria was formed by rapid quenching from temperatures higher than 1500°C.

Ryu et al. [2010b] used pre-mixed zircona-yttria precursors to investigate the formation of yttria-stabilized tetragonal phase in the plasma process described above. The precursors were ground and mixed in a jar mill. Tests were performed under the conditions of the plasma torch power of 13 kW, the plasma gas flow rate of 39 L/min (25°C, 86.1 kPa), the feeding rate of premixed precursor of 2 g/min, and the flow rate of Ar to carry the precursor powder of 4 L/min (25°C, 86.1 kPa).

Figure 8.5 shows the XRD patterns of the products obtained with Y_2O_3 contents aimed to produce 4 mol% Y_2O_3-ZrO_2 (referred to as "4YSZ") and 8 mol% Y_2O_3-ZrO_2 (referred to as "8YSZ"). The XRD patterns of the products shown in Figures 8-5(a) and (b) correspond to those of the tetragonal phase zirconia with no monoclinic phase zirconia detected. (Major monoclinic peaks occur at $2\theta = 28.1$, 31.4, 34.1 and 50.1.) Because the overall XRD patterns of tetragonal and cubic zirconia are similar, it is difficult to distinguish between these two phases. However, it has been established that the cubic phase can be identified from the diffraction patterns in the 58°-62° range and the 72°-75° range [Cullity, 1978; Gibson and Irvine, 2001]. Considering this, 8YSZ also displayed features of cubic phase.

To more clearly identify the phases of ZrO_2 in the product, Raman spectroscopy analysis was conducted as shown in Figure 8.6. The Raman spectrum for the tetragonal phase is characterized by six sharp peaks at approximately 146, 261, 318, 472, 613, and 636 cm^{-1}, while the cubic phase only shows the spectrum of a single broad peak in the range of 530-670 cm^{-1} [Chervin et al., 2006; Li and Li, 2002]. The Raman spectrum of 4YSZ shown in Figure 8.6(a) clearly exhibits peaks corresponding to the tetragonal phase. The 8YSZ spectrum shown in Figure 8.6(b) also displays peaks for the tetragonal phase, but the peaks are rather broad and their intensities are low, which indicates less tetragonal characteristics than the 4YSZ in Figure 8.6(a) [Chervin et al., 2006; Li and Li, 2002]. Also, the broad peak near 646 cm^{-1} observed in Figure 8.6(b) could be due to cubic phase zirconia in the product. Based on the XRD and Raman spectroscopy results, it was concluded that 4YSZ and 8YSZ consisted, respectively, of pure tetragonal phase zirconia and a mixture of tetragonal and cubic phases of zirconia.

EDX analysis was also performed, to confirm the presence of yttria in 4YSZ and 8YSZ, as shown in Figure 8.7. The yttria contents in 4YSZ and 8YSZ were 3.8 mol% and 7.7 mol%, respectively. It has been reported that the amount of yttria-doping affects the zirconia phase [Chervin et al., 2006; Yashima et al., 1996]. Chervin et al. [2006] investigated the phase transition of yttria-stabilized zirconia over the range of 3-25 mol% yttria. The tetragonal phase was predominantly formed at 3.07 mol% yttria, whereas a mixture of tetragonal and cubic was observed at 7.69 mol%. These results are in good agreement with this study, although the synthesis methods used were different.

The grain sizes of 4YSZ and 8YSZ calculated by the Scherrer equation were about 21 nm independent of the amount of yttria doping. The particle sizes were also examined using ZetaPALS and TEM. The ZetaPALS measurements yielded average particle sizes of 88 nm and 93 nm, respectively. However, TEM examinations resulted in average particle sizes of both samples of 27 nm with a size range from 10 nm to 50 nm, as shown in Figure 8.8. The larger particle size obtained by ZetaPALS is likely due to the agglomeration of particles in liquid media.

The specific surface areas of 4YSZ and 8YSZ powders measured by a BET method were 29 ± 0.9 m^2/g and 28 ± 0.8 m^2/g, respectively. From the results above, it was determined that the particle size and surface area were nearly constant, irrespective of the amount of yttria doping.

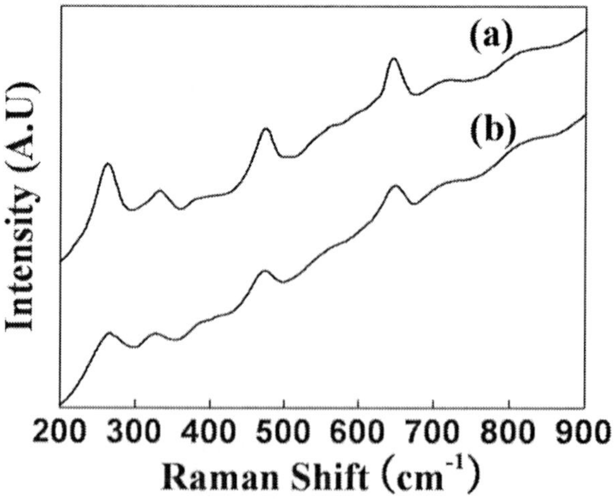

Figure 8.6. Raman spectra of (a) 4YSZ and (b) 8YSZ. [Adapted from Ryu et al. (2010b)].

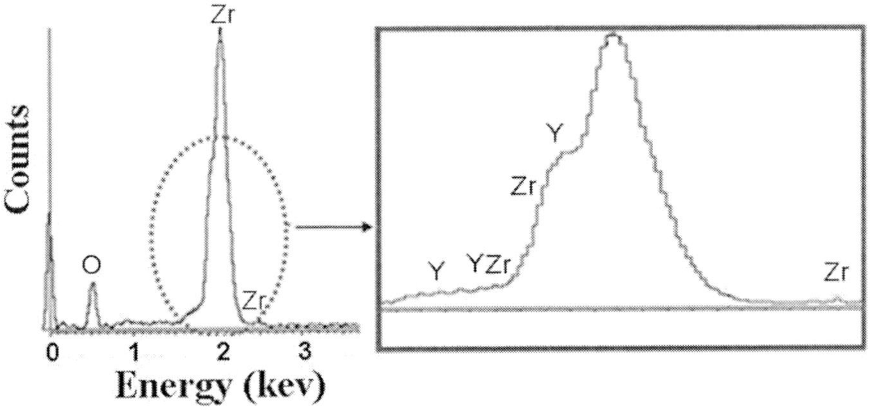

Figure 8.7. EDX spectrum for the synthesized 8YSZ powder. [Adapted from Ryu et al. (2010b)].

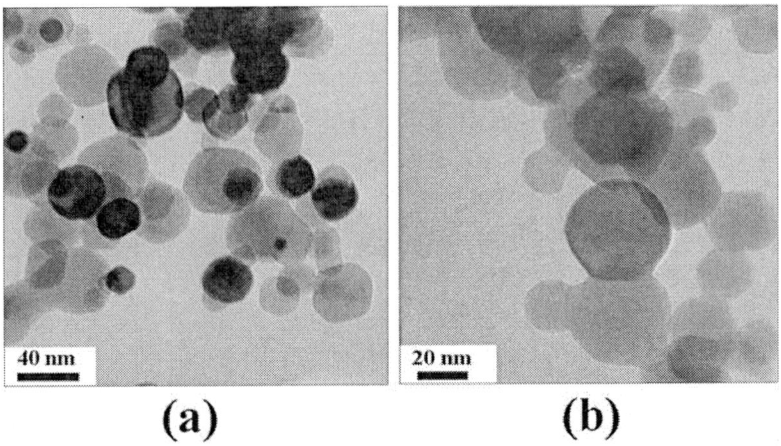

Figure 8.8. TEM micrographs of (a) 4YSZ and (b) 8YSZ. [Adapted from Ryu et al. (2010b)].

8.2 Thermal Stability of Yttria-Stabilized Zirconia Nanopowder

Figure 8.9 shows the DTA results of yttria-doped zirconia (4YSZ and 8YSZ) and non-doped zirconia (tetragonal phase zirconia referred to as "t-ZrO_2"). The samples were heated in Ar from room temperature to 1400°C at a heating rate of 10°C/min, held at 1400°C for 1 hour, and cooled to room temperature at a cooling rate of 10°C/min. Undoped zirconia exhibited an exothermic peak around 1030°C on cooling, as shown Figure 8.9(a), whereas no exothermic or endothermic peaks were observed in yttria-doped zirconia, as shown in Figures 8-9(b) and (c).

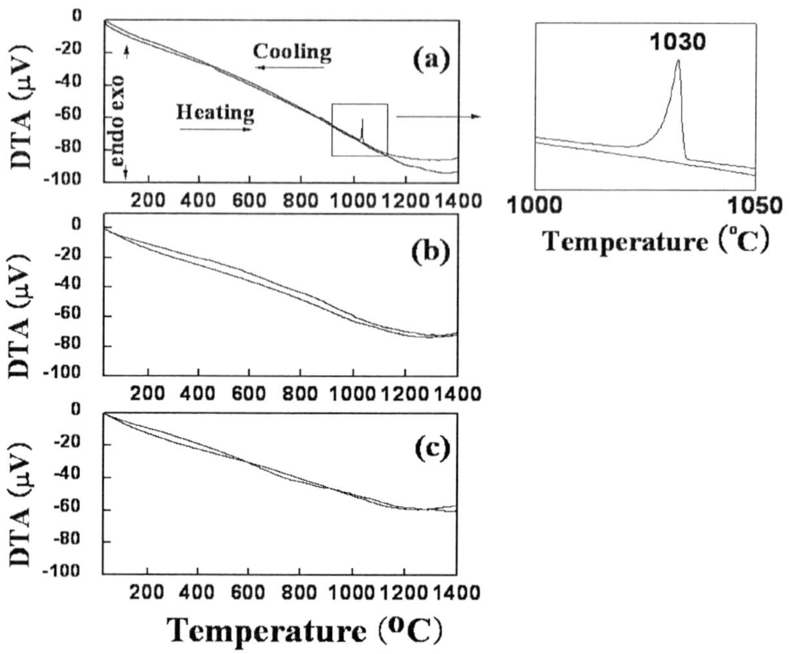

Figure 8.9. DTA patterns of (a) t-ZrO_2, (b) 4YSZ, and (c) 8YSZ. [In (a) the starting tetragonal phase apparently does not transform to monoclinic phase on heating at 5°C/min heating rate used in this work, based on the absence of an endothermic peak around 1200°C.] [Adapted from Ryu et al. (2010b)].

It has been well known that the phase transition of undoped ZrO_2 from monoclinic to tetragonal phase occurs at around 1200°C on heating and from tetragonal to monoclinic phase is observed at around 1000°C on cooling [Tartaj et al., 2007; Kiminami, 1990; Ruh et al. 1984]. Based on this, the exothermic peak observed in Figure 8.9(a) was attributed to the phase transition from tetragonal to monoclinic on cooling, while the lack of an exothermic peak in Figures 8-9(b) and (c) indicated that phase transition did not occur. In other words, the yttria-doped zirconia (4YSZ and 8YSZ) synthesized by the plasma process method is stable at high temperatures without the monoclinic-tetragonal phase transition.

To summarize, nano-sized zirconia and yttria-stabilized zirconia powders were synthesized by a thermal plasma process with zirconium carbonate and/or yttrium carbonate as the starting materials. During the synthesis of zirconia, a mixture of monoclinic and tetragonal phases was formed. As the plasma torch power or plasma gas flow rate increased, the amount of tetragonal phase in the mixture increased. The average particle size of zirconia was 17 nm when formed at 7 kW, and increased to 23 nm with an increase of plasma torch power to 34 kW.

Pure tetragonal phase zirconia was formed at 3.8 mol% yttria, while a mixture of tetragonal and cubic phases was observed at 7.7 mol% yttria. The particle sizes of both samples were approximately 27 nm. Tests for thermal stability indicated that the yttria-doped tetragonal phase was stable and did not exhibit any phase transition, while non-doped tetragonal phase displayed tetragonal-monoclinic phase transition around 1030°C during cooling.

The thermal plasma process performed in this work has shown its potential as an efficient technique for preparing nano-sized oxide powders. This process is also suitable for large-scale production of nano-sized powders owing to the availability of high temperatures for rapidly volatilizing reactants followed by vapor phase reaction and rapid quenching of the product to directly yield nano-sized powders.

Chapter 9

COMPUTATIONAL FLUID DYNAMIC MODELING OF THE-CHEMICAL VAPOR SYNTHESIS PROCESSES

ABSTRACT

Chemical vapor synthesis (CVS) processes involve chemical reactions in a flowing gas stream accompanied by a number of complex phenomena, including the nucleation and growth of a condensed phase. The final particle size, composition and morphology of the powders are determined by the combined effects of these factors. It would appear that mathematically modeling this complex process is a daunting task. However, there has been a significant progress made on the topic in recent years, which will be described in this chapter. In fact, because the process is so complicated, a mathematical model based on fundamental principles provides a critical tool in the study and application of the CVS processes.

NOMENCLATURE

A_d	Spray droplet surface area (m^2)
A_p	Total surface area of particles per unit volume of gas-particle mixture (m^2 m^{-3})
$\vec{A}$	Cross-sectional area vector of the reactor outlet (m^2)
$\bar{c}_i$	Mean molecular velocity of species i defined by Equation (9-18); $\bar{c}_i = \left(\frac{8k_BT}{\pi m_i}\right)^{1/2}$ (m s^{-1})
$\bar{c}_j$	Mean thermal velocity of particle j defined by Equation (9-22); $\bar{c}_j = \left(\frac{8k_BT}{\pi m_j}\right)^{1/2}$ (m s^{-1})
$c_{p,d}$	Droplet heat capacity at constant pressure (J kg^{-1} K^{-1})
C_c	Slip correction factor defined by Equation (9-21);

$$C_c = 1 + \frac{2\lambda}{d_{p,j}}\left(1.257 + 0.4\exp\left(-\frac{1.1 d_{p,j}}{2\lambda}\right)\right) \text{(dimensionless)}$$

C_D	Drag coefficient (dimensionless)
$C_{i,b}$	Molar concentration of species i in the bulk (mol m^{-3})
$C_{i,s}$	Molar concentration of species i at the droplet surface (mol m^{-3})
c.v.	Local coefficient of variation (dimensionless)
d_d	Spray droplet diameter (m)
d_p	Particle size (diameter) (m)
d_p^*	Critical particle diameter (m)
$\bar{d}_{p,exp}$	Experimentally determined average particle diameter (m)
d_{pg}	Geometric mean diameter (m)
$d_{p,Sauter}$	Sauter mean diameter defined by Equation (9-48); $d_{p,Sauter} = m_3 / m_2$ (m)
$\bar{d}_{p,Sauter}$	Sauter mean diameter of all the particles leaving the reactor defined by Equation (9-50); $\bar{d}_{p,Sauter} = \dot{M}_3 / \dot{M}_2$ (m)
D_{i-mix}	Diffusion coefficient of species i in the vapor mixture (m^2 s^{-1})
D_j	Diffusion coefficient of particle j defined by Equation (9-20); $D_j = \frac{k_B T C_c}{3\pi\mu d_{p,j}}$ (m^2 s^{-1})
f_k	Volume fraction of phase k (m^3 m^{-3})
g_j	Transition parameter of particle j defined by Equation (9-23); $g_j = \frac{1}{3 d_{p,j} l_j}\left[\left(d_{p,j} + l_j\right)^3 - \left(d_{p,j}^2 + l_j^2\right)^{3/2}\right] - d_{p,j}$ (m)
G	Linear particle growth rate by vapor condensation (gas-to-particle conversion) (m s^{-1})
G_c	Linear particle growth rate under the control of chemical reaction kinetics (m s^{-1})
G_m	Linear particle growth rate under the control of mass transfer (m s^{-1})
$G_{overall}$	Overall linear particle growth rate under the combined control of surface chemical reaction and mass transfer (m s^{-1})
h	Heat transfer coefficient (J m^{-2} K^{-1})
$h_{fg,i}$	Vaporization enthalpy of species i (J kg^{-1})
J	Particle nucleation rate (# s^{-1} m^{-3})
k_B	Boltzmann constant (=1.38x10^{23} J K^{-1})
$k_{c,i}$	Convective mass transfer coefficient of species i (m s^{-1})
k_d	Decomposition rate constant (s^{-1})
k_g	Growth rate constant used in Equation (9-39); (m^7·mol^{-2}·s^{-1})

k_n	Nucleation rate constant used in Equation (9-38); ($m^3 \cdot mol^{-1} \cdot s^{-1}$)
Kn	Knudsen number defined by Equation (9-17); $\mathrm{Kn} = \frac{6D_{i-mix}}{\bar{c}_i d_p}$ (dimensionless)
l_j	Mean free path of particle j defined by Equation (9-24); $l_j = \frac{8D_j}{\pi \bar{c}_j}$ (m)
m_i	Molecular mass of vapor species i (kg)
m_d	Spray droplet mass (kg)
$m_{i,d}$	Mass of species i in the spray droplet (kg)
M_k	k-th moment of the particle size distribution ($m^k\ m^{-3}$)
$\dot{M}_v$	Flow rate of the quantity related to k-th moment, defined in Equation (9-51) ($m^k\ s^{-1}$)
m_j	Mass of particle j (kg)
M_i	Molecular weight of species i (kg mol^{-1})
n	Particle number density function (# $m^{-3}\ m^{-1}$)
N	Avogadro's number (# mol^{-1})
N_A	Number concentration of species A in the vapor phase (# m^{-3})
p_i	Partial pressure of species i (Pa)
p_i°	Vapor pressure of species i at the particle surface (Pa)
r	Radial position (m)
R	Gas constant (J $mol^{-1}\ K^{-1}$)
S	Supersaturation ratio defined by Equation (9-14); $S = \frac{p_i}{p_i^\circ}$ (dimensionless)
S_p	Computed local specific surface area of the particles ($m^2\ kg^{-1}$)
$S_{p,exp}$	Measured specific surface area ($m^2\ kg^{-1}$)
S_ϕ	Net source term of dependent variable ϕ (various)
Sh	Sherwood number defined by Equation (9-42); $\mathrm{Sh} = \frac{k_{c,i} d_p}{D_{i-mix}}$ (dimensionless)
T	Temperature (K)
T_d	Spray droplet temperature (K)
$\vec{u}$	Spray droplet velocity vector (m s^{-1})
v_i	Molecular volume of species i (m^3)
$\vec{v}_g$	Gas velocity vector (m s^{-1})
v_p	Particle volume (m^3)
$\vec{v}_p$	Particle-phase velocity vector (m s^{-1})
$\vec{v}_m$	Mixture velocity vector (m s^{-1})

$v_{m,r}$	Radial component of the mixture velocity vector (m s^{-1})
$v_{m,z}$	Axial component of the mixture velocity vector (m s^{-1})
V_A	Molar volume of species A (m^3 mol^{-1})
V_p	Total volume of particles per unit volume of gas-particle mixture (dimensionless)
Y_i	Mass fraction of species i (dimensionless)
z	Axial position (m)

Greek Symbols

α_v	Vapor-phase collision efficiency factor (dimensionless)
α_p	Particle collision efficiency factor (dimensionless)
β	Collision frequency function (m^3 s^{-1})
Γ_{ϕ_m}	Mixture effective transport coefficient (kg m^{-1} s^{-1}, W m^{-1} K^{-1}, or m^2 s^{-1})
λ	Mean free path of the vapor molecules (m)
μ	Gas viscosity (Pa s)
ρ_d	Spray droplet density defined by Equation (9-36); $\rho_d = \left(\sum_i \frac{m_{i,d}}{m_d \rho_{i,d}}\right)^{-1}$ (kg m^{-3})
$\rho_{i,d}$	Mass density of species i (kg m^{-3})
ρ_m	Mixture density defined by Equation (9-3); $\rho_m = \sum_{k=1}^{K} f_k \rho_k$ (kg m^{-3})
ρ_M	Molar density (mol·m^{-3})
ρ_p	Particle-phase density (kg m^{-3})
σ	Surface tension (N m^{-1})
σ_{gs}	Geometric standard deviation defined by Equation (9-27); $\ln^2 \sigma_{gs} = \ln \frac{m_0 m_2}{m_1^2}$
ϕ_m	Mixture generic dependent variable (various)

Subscripts

A	Condensing vapor species
b	Bulk
d	Spray droplet
i	Species i
j	Particle j
k	Phase k, k-th moment
m	Mixture
p	Particle
s	Spray droplet surface

Operators

∇ Nabla operator in Cartesian coordinates; $\nabla = \frac{\partial}{\partial x}\delta_x + \frac{\partial}{\partial y}\delta_y + \frac{\partial}{\partial z}\delta_z$ (m^{-1})

9.1. INTRODUCTION

A process model consists of process steps of the same nature as common blocks. Thus, a process model is a description of a process as an abstract combination of these blocks and can be used to anticipate what the actual process will look like. Since we want to obtain quantitative results from the model, the behavior of the block is typically described by mathematical equations representing the process taking place together with known properties and knowledge about the process. The purposes of process models can be to describe, explain, and/or predict what happens in a process. The major individual objectives include: to increase fundamental understanding of a process; to assist in scale-up; and to be used in process control and optimization.

Process models track elemental, compound and energy components of a process. Material and energy balance calculations in chemical vapor synthesis (CVS) processes can be very complex. The development of a process model starts with the definition of what aspects of the process one wants to simulate. This is followed by the determination of which phenomena must be included and which to simplify or ignore. Next, the type and complexity of the model must be chosen. The purpose of the model should dictate what phenomena are included, and consequently how complex the model should be. The next step is to model phenomena in proportion both to their importance to the problem and to how well they are understood mechanistically. Important phenomena that are well understood should be included in detail. Overly complex models are too difficult and time consuming to run. On the other hand, unreasonable simplifications lead to erroneous conclusions.

Mathematical equations that make up a process model are obtained by applying the laws of mass, energy, and momentum conservation. A macroscopic model may describe the average temperature, average composition, and average velocity in a reactor, but cannot describe how these properties vary with position within the reactor. Thus, important questions such as how large the unit must be or whether conditions for a local hot spot or excessive erosion may be created cannot be correctly answered by such a model. To be able to describe these details, the conservation laws must be applied to a very small portion of the unit – the smaller the control volume, the more accurate the results. Thus, a microscopic model that can describe the continuous variation of the system properties is based on an infinitesimally small control volume, usually giving rise to differential equations [Bird et al., 2002]. A balance is performed on the quantity (mass, heat or momentum) to be conserved over such a control volume, as follows:

$$\begin{bmatrix} \textit{Input across} \\ \textit{boundary of} \\ \textit{control volume} \end{bmatrix} - \begin{bmatrix} \textit{Output across} \\ \textit{boundary of} \\ \textit{control volume} \end{bmatrix} + \begin{bmatrix} \textit{Net generation} \\ \textit{within} \\ \textit{control volume} \end{bmatrix} = \begin{bmatrix} \textit{Accumulation} \\ \textit{within} \\ \textit{control volume} \end{bmatrix} \tag{9-1}$$

These balance equations yield the governing equations for concentration, temperature, and velocity in terms of position and time. The solution of these governing equations has become convenient with the availability of many powerful commercial software packages.

The results of the model predictions, especially for problems with complex model equations such as those involving fluid flow, must be verified by testing the program under appropriate conditions that will yield known solutions. Once the process model is verified, its predictions should be compared with experiments and any disagreements must be reconciled either by modifying the model equations or by repeating the experiments. Reliable results can only be obtained by careful examination of input parameters/assumptions and laboratory or operating data.

The chemical vapor synthesis (CVS) processes for nanopowder production involve chemical reactions in a flowing gas stream accompanied by the nucleation and growth of a condensed phase; the transfer of gaseous species within the gas phase as well as between the bulk gas and the condensed phase; and usually transfer of heat within and between the phases. In these processes, the final particle size, composition and morphology of the powders are determined by the reaction temperature, residence time of the particle in the reaction zone, position of reactant feeders, and fluid dynamic conditions inside the reactor. Although much experimental work has been done on vapor-phase processes, the mechanisms of formation, growth, and transport of nanoparticles in high-temperature reactors with fast chemical reactions and complex fluid dynamics are still not well understood. As a result, the experimental design and scale-up of this process is a rather difficult task. Thus, mathematical models based on fundamental principles will provide a critical tool in the study and application of vapor-phase processes. Because of the complexity involved, mathematical modeling of CVS processes typically requires the use of Computational Fluid Dynamics (CFD) techniques.

A CVS reactor usually has a complicated internal structure and flow configuration involving mixing of a number of reactant streams, and thus a comprehensive CFD model of CVS needs to solve the three-dimensional turbulent governing equations of overall continuity, momentum, energy, species transport as well as chemical kinetics in a multiphase domain. The description of the condensed phase requires obtaining local particle size distribution (PSD) by solving the population balance model. The nucleation of the condensed phase involves different controlling phenomena depending on the nature of the nucleation process. A brief review of the homogeneous nucleation theory will be presented and main factors affecting the nucleation rate of metal and ceramic particles will be discussed. The growth of the particles takes place by gas-to-particle conversion (vapor condensation and surface reaction), Brownian coagulation, or a combination thereof, depending on the system. The numerical solution is typically obtained by the use of a commercial computational fluid dynamics (CFD) code adapted to incorporate the nucleation and growth phenomena.

The model predictions are compared with experimental results typically in terms of the average particle size and other process parameters. The effects of the operating conditions such as reaction temperature and carrier gas feed rate on the PSD are evaluated. The use of computational models to simulate nanopowder synthesis reactors [Zucca et al., 2005; Schwade and Roth, 2003] as an engineering design tool is gaining importance due to the increased availability of faster computers. In this chapter, CFD modeling performed on a number of CVS systems for nanopowder preparation is discussed.

9.2. Model Formulation

The mathematical modeling of various processes involved in the vapor-phase synthesis of nanopowders requires solution of the coupled gas-phase transport equations, i.e. fluid flow, heat and mass transfer, chemical reaction kinetics, and the particle population balance model (PBM). The particle-laden (multiphase) flow is described by the mixture model [ANSYS, 2006] in which the governing equations for the vapor and particulate phases are treated. The model formulation includes the following two general assumptions: (a) steady state and (b) negligible effect of the particles on the aerodynamic behavior of the gas phase in simulating a CVS process because of the very fine particles size.

Governing Equations

The mixture model solves the governing equations for the mixture (vapor and particles) and the continuity equation for the particulate phase. The generic mixture transport equation, in steady state, can be written in the following form:

$$\nabla \cdot \rho_m \vec{v}_m \phi_m - \nabla \cdot \Gamma_{\phi,m} \nabla \phi_m = S_{\phi,m} . \tag{9-2}$$

The density of the mixture is given by

$$\rho_m = \sum_{k=1}^{n} f_k \rho_k \tag{9-3}$$

where f_k is the volume fraction of phase k and ρ_k is the density of phase k. In general, any mixture-phase dependent variable is computed by

$$\phi_m = \sum_{k=1}^{n} f_k \phi_k . \tag{9-4}$$

The mixture-phase equations take on the same form as the gas-phase transport equations which can be found in standard textbooks on transport phenomena [e.g. Bird et al., 2002].

The continuity equation for the particulate phase is given by

$$\nabla \cdot f_p \rho_p \vec{v}_p = S_p \tag{9-5}$$

where f_p is the particle volume fraction. The term on the right-hand side of Equation (9-5) represents the mass addition from the gas phase to the particulate phase, i.e. particle nucleation and growth. Since it is assumed that there is a negligible effect of the particles on the aerodynamic behavior of the gas phase, the velocity vectors of the mixture, gas phase and particles are all equal ($\vec{v}_m = \vec{v}_g = \vec{v}_p$). This assumption can be relaxed by incorporating the appropriate laws for the motion of particles in a fluid environment [Friedlander, 2000].

A total of $n + 8$ transport equations of the form of Equation (9-2) are needed to describe the mixture-phase processes, i.e. overall continuity, momentum in the three Cartesian directions, turbulent kinetic energy, turbulent dissipation rate, energy, radiation intensity, n - 1 (of n total) species continuity equations, and the continuity equation for the particles. The gas phase is typically assumed to obey the ideal gas law. The standard k-ε turbulence model is employed. The heat transfer by radiation is described using the discrete ordinate (DO) model [ANSYS, 2006].

Gas-Phase Reaction Kinetics

The vapor-phase synthesis of nanopowders starts with the mixing and reaction of the vapor precursors to form condensable products. The gas-phase reaction can be represented by the following overall equation:

$$aA(g) + bB(g) = cC(g) + dD(g). \tag{9-6}$$

The produced vapor C is then condensed by particle nucleation and growth as follows

$$C(g) = C(s,l) \tag{9-7}$$

where C(s,l) is the condensed phase.

In the cases where the synthesis process is conducted at high temperature and the mixing of the vapor precursors takes place inside the reactor; it is customary to assume that the intrinsic reaction rate is very fast and thus the reaction rate is controlled by the turbulent mixing of the vapor reactants. As a result, the rate of reaction of Equation (9-6) can be calculated using the Eddy-Dissipation Model (EDM) [Magnussen and Hjertager, 1976].

Population Balance Model

The population balance model (PBM), which incorporates particle transport, nucleation, growth and coagulation, describes the change in particle size distribution function in terms of the particle size and the local spatial position [Friedlander, 2000]. At steady state, the PBM is expressed by the following equation:

$$\nabla \cdot n\vec{v}_m + \frac{\partial}{\partial d_p}(nG) = B_{coag} - D_{coag} \tag{9-8}$$

where n is the particle number density function, d_p is the particle size (diameter), and G is the particle size growth rate. The first term on the left-hand side of Equation (9-8) denotes the convection of particles due to the flow field. The second term represents the growth of particles by gas-to-particle conversion, i.e. nucleation, vapor condensation, and surface reaction. The terms on the right-hand side describe particle coagulation and are given by

$$B_{coag} = \frac{1}{2}\int_0^{d_p} \beta(d_p', (d_p^3 - d_p'^3)^{1/3}) n(d_p') n((d_p^3 - d_p'^3)^{1/3}) dd_p' \tag{9-9}$$

and

$$D_{coag} = n(d_p)\int_0^{\infty} \beta(d_p', d_p) n(d_p') dd_p' . \tag{9-10}$$

Equation (9-9) represents the rate of birth of particles of size d_p by collisions between particles of sizes $d_p - d_p'$ and d_p'. The term $1/2$ prevents counting the collisions twice, once each for the two colliding particles. Equation (9-10) describes the rate of death of particles of size d_p by collisions of particle of size d_p with particles of all sizes.

The boundary condition at $d_p = 0$ for the integration of the second term on the left-hand side of Equation (9-8) is given by

$$n\big|_{d_p=0} = \frac{J}{G} \tag{9-11}$$

where J represents the rate of nucleation of particles.

The particle size distribution function can be expressed in terms of the appropriate moments of the size distribution. The k-th moment M_k is defined by [Friedlander, 2000]

$$M_k = \int_0^{\infty} d_p^k n(d_p) dd_p \tag{9-12}$$

The zeroth moment (m_o) is equal to the total number of particles per unit volume of the system. The first moment (m_1) gives the sum of the particle diameters per unit volume. The second moment (m_2) gives, when multiplied by π, the total surface area of particles per unit volume. The third moment (m_3) yields, when multiplied by π/6, the total volume of particles per unit volume.

Kinetics of Particle Nucleation and Growth

To solve the PBM given by Equation (9-8), expressions for the rates of particle nucleation, growth and coagulation must be provided. In this section, these expressions will be introduced along with the underlying theories.

Particle Nucleation: Homogeneous nucleation is responsible for the formation of particles from the vapor phase. In this process, vapor atoms or molecules are transformed into solid particles in a supersaturated vapor without the presence of a foreign phase. In chemical vapor synthesis where vapor-phase reactions produce the condensable product, the homogeneous nucleation process can be controlled by the degree of supersaturation [Seinfeld and Pandis, 2006] or it can be reaction-limited [Kodas and Hampden-Smith, 1999; Bolsaitis et al., 1987] depending on the vapor pressure of the condensed phase.

The classical theory of homogeneous nucleation describes the formation of nuclei when the degree of supersaturation of the vapor phase is the driving force. In a supersaturated vapor, small clusters of the secondary phase containing a few molecules are produced. However, most of them are thermodynamically unstable and disappear in a short time. Only a small fraction of these clusters will grow to a size large enough to produce nuclei, i.e. the clusters have reached their critical size. In a cluster of critical size, molecules condense and evaporate at the same rate. Hence, those clusters that have a size larger than the critical size will probably continue to grow and become particles. In contrast, clusters with a size smaller than the critical size will evaporate. The critical nucleus size is predicted by the Kelvin equation [Friedlander, 2000] given by

$$d_p^* = \frac{2\sigma}{\rho_M RT \ln S} \tag{9-13}$$

where the supersaturation ratio S is defined as follows

$$S = \frac{p_i}{p_i^\circ}. \tag{9-14}$$

The nucleation rate is therefore the net number of clusters formed per unit time that has a size larger than the critical size and can be computed by [Seinfeld and Pandis, 2006]

$$J = \left(\frac{2\sigma}{\pi m_i}\right)^{1/2} \left(\frac{v_i N_i^2}{S}\right) \exp\left(-\frac{16\pi}{3} \frac{v_i^2 \sigma^3}{(k_B T)^3 (\ln S)^2}\right). \tag{9-15}$$

Reaction-controlled nucleation occurs when the vapor pressure of the nucleating species is very low [Kodas and Hampden-Smith, 1999; Rao and McMurry, 1989; Bolsaitis et al., 1987]. Under such conditions, the nucleation process is indistinguishable from the chemical reactions that produce the condensable products. In other words, every monomer formed by chemical reaction in the vapor phase can be regarded as a thermodynamically stable particle. Thus, the nucleation rate can be represented by the rate of formation of the vapor product.

Particle Growth by Gas-to-Particle Conversion: In vapor-phase synthesis, the growth of particles by gas-to-particle conversion occurs as the result of vapor condensation and/or surface reactions. In vapor condensation, particles grow when vapor atoms or molecules condense on the particle surface. The rate of change of the particle diameter with time by condensation is computed by the Fuchs-Sutugin interpolation formula [Seinfeld and Pandis, 2006; Friedlander, 2000], given by

$$G = \frac{dd_p}{dt} = \frac{4V_i D_{i\text{-}mix}}{d_p RT}\left(p_{i,b} - p_i^{\circ}\right)\left\{\frac{0.75\alpha_v(1+\text{Kn})}{\text{Kn}^2 + \text{Kn} + 0.283\text{Kn}\,\alpha_v + 0.75\alpha_v}\right\} \qquad (9\text{-}16)$$

where α_v is the vapor collision efficiency factor, Kn is the Knudsen number defined as

$$\text{Kn} = \frac{6D_{i\text{-}mix}}{\bar{c}_i d_p} \qquad (9\text{-}17)$$

and $\bar{c}_i$ is the mean molecular velocity of species *i* expressed as follows

$$\bar{c}_i = \left(\frac{8k_B T}{\pi m_i}\right)^{1/2}. \qquad (9\text{-}18)$$

Equation (9-16) represents the growth rate in the entire range of Knudsen numbers. When $\text{Kn} << 1$, Equation (9-16) becomes the diffusion equation for the continuum regime. When $\text{Kn} >> 1$, Equation (9-16) becomes the growth equation in the free-molecule regime.

In certain systems, chemical reactions on the particle surface are the important mechanism for particle growth. This mechanism is known as surface reaction growth and was used in the simulation of tungsten carbide nanopowders. Further details on the modeling equations will be provided in Section 9.5.

Particle Coagulation: The coagulation of nanoparticles takes place predominantly by Brownian motion. In this mechanism, two particles collide and stick together to form a new, larger particle. Coagulation increases the average particle size, decreases the total particle number density, and conserves the total particle mass concentration and volume [Friedlander, 2000]. Particle coagulation is often modeled by assuming that the rate of particle coalescence is infinitely fast, allowing the particle to always be spherical. The collision frequency function is computed using the Fuchs interpolation formula given by [Seinfeld and Pandis, 2006]

(9-19)

where D_j is the diffusion coefficient of particle *j* given by

$$D_j = \frac{k_B T C_c}{3\pi\mu d_{p,j}}; \qquad (9\text{-}20)$$

is the slip correction factor defined by

$$C_c = 1 + \frac{2\lambda}{d_{p,j}}\left(1.257 + 0.4\exp\left(-\frac{1.1d_{p,j}}{2\lambda}\right)\right); \qquad (9\text{-}21)$$

is the mean thermal velocity of particle j expressed by the following equation

$$\bar{c}_j = \left(\frac{8k_B T}{\pi m_j} \right)^{1/2}; \tag{9-22}$$

is the transition parameter given by

$$g_j = \frac{1}{3d_{p,j} l_j} \left[\left(d_{p,j} + l_j \right)^3 - \left(d_{p,j}^2 + l_j^2 \right)^{3/2} \right] - d_{p,j}; \tag{9-23}$$

denotes the mean free path of particle j as follows:

$$l_j = \frac{8D_j}{\pi \bar{c}_j}; \tag{9-24}$$

and α_p is the particle collision efficiency factor. Equation (9-19) allows the calculation of the collision frequency function from the free-molecule regime ($\mathrm{Kn} >> 1$) to the continuum regime ($\mathrm{Kn} << 1$), including the transition ($\mathrm{Kn} \approx 1$) stage. The parameter g_j is related to the average distance that a particle travels in the transition regime ($\mathrm{Kn} \approx 1$) before colliding with another particle.

Numerical Solution

The numerical solution of the transport equations (Equations (9-2) and (9-5)) subject to the appropriate boundary conditions is obtained by the use of the Fluent CFD code [ANSYS, 2006]. The reactor geometries are discretized on non-uniform grids using a finite volume approach. The PBM (Equation (9-8)) is solved by means of the quadrature method of moments (QMOM) [McGraw, 1997], which allows the direct tracking of the local particle size distribution (PSD).

The techniques described above have been used to simulate the synthesis of aluminum (Al), silica (SiO_2), and tungsten carbide (WC) nanopowders from the vapor-phase reactions of precursors. These examples will be discussed in the following sections.

9.3. Simulation of Aluminum Nanopowder CVS

The importance of aluminum nanopowder in various industrial fields and the CVS process of its preparation have been discussed in Section 3.3. Several researchers [Setyawan and Yuwana, 2008; Prakash et al., 2003; Schefflan et al., 2001; Panda and Pratsinis, 1995] have worked on the modeling of the synthesis processes of aluminum particles. Most of these efforts have been focused on solving the population balance model in highly idealized flow conditions, such as plug-flow reactors. Thus, the effects of the fluid flow, heat and mass transfer and chemical kinetics on the particle size distribution were not incorporated. This simplified approach is suitable when the synthesis reactor has a simple geometrical configuration such as a tubular reactor. However, this type of reactors is rarely encountered in practice, making mandatory the use of computational fluid dynamics (CFD) when modeling real reactors. With this in mind, Sohn et al. [2010] proposed a CFD model for a CVS reactor for the synthesis of aluminum nanopowder. Although the simulated reactor is at the laboratory scale, it presents a complex geometrical configuration which requires a full 3-D simulation. In this work, the rates of nucleation and growth were computed using greatly simplified expressions. Since no experimental data are available on these rates, the kinetic parameters for nucleation and growth were treated as adjustable parameters.

Here, the results of the three-dimensional simulation by Perez-Fontes and Sohn [2011] of a laboratory-scale CVS reactor for the synthesis of aluminum nanopowder, using the comprehensive approach described above, are presented. A schematic representation of the reactor used for this work was shown in Figure 3.10 of Section 3.3. It consists mainly of a horizontal tubular reactor, an electrically heated furnace, two separate powder feeders, and evaporators. $AlCl_3$ and Mg powders are delivered into the reactor by specially designed powder feeders. The fed $AlCl_3$ and Mg powders are evaporated in two separate evaporators placed inside the reactor at the same axial position. Argon gas is supplied to control the concentration and residence time of the reactants and to prevent back flow. The formation of aluminum proceeds according to

$$AlCl_3(g) + 3/2Mg(g) = Al(s) + 3/2MgCl_2(g) \quad (9\text{-}25)$$

The produced aluminum powder is collected in a powder collector outside the reactor.

Simulation Strategy: The model described above was applied to the formation of aluminum nanopowder from the chemical reaction between the gaseous reactants $AlCl_3$ and Mg [Perez-Fontes and Sohn, 2011].

Kinetics of Particle Formation and Growth: The Kelvin equation, Equation (9-13), predicts that the critical nucleus size of an aluminum particle is larger than its monomer size under this experimental condition. Therefore, the nucleation rate of aluminum particles was computed using the expression in Equation (9-15), based on the classical nucleation theory.

Particle growth was modeled by vapor condensation [Equation (9-16)] and Brownian coagulation [Equation (9-19)]. The collision efficiency factor α for aluminum particles was set equal to unity based on the fact that the reaction temperature is higher than the melting point of aluminum. This value yielded particle sizes that agreed reasonably well with the experimental measurements [Perez-Fontes and Sohn, 2011].

Computational Mesh and Simulation Conditions: Figure 9.1 shows the geometry of the simulated reactor and the computational mesh, and Figure 9.2 illustrates the isometric view of the computational geometry. In order to minimize the computational time, the geometry was divided in two regions, namely, Region I and Region II. Then, 3-D meshes were made using

Gambit 2.2.30 for each region. The solution using Region I, which considered only the first part of the reactor, was used to obtain profiles of the dependent variables at a plane perpendicular to the flow direction at an axial position (0.6 m) before the end of the injection tubes. At this axial position, particles have not formed yet and thus only the gas-phase problem was solved. The obtained profiles which included temperature, velocity components, species mass fraction of Mg, $AlCl_3$, $MgCl_2$, Al and Ar, turbulence kinetic energy and turbulent dissipation rate were used as inlet boundary conditions for the simulation of Region II. With this geometry the full multiphase problem was solved.

It was shown that the reactor zone which is located outside the furnace on the right-hand side of Figure 9.1a is particularly important for particle coagulation.

Simulation runs with a reduced reactor length resulted in smaller particle sizes. The experimental temperature profile given in Sohn et al. [2010] was used as a boundary condition at the furnace wall. The left-hand side of Region II (Figure 9.1b) represents the inlet where the profiles were included, while the right-hand side represents the reactor outlet. The total number of cells for the simulation was 385,076. The cross-sectional view of the mesh in Section A is shown in Figure 9.1c.

Velocity, Temperature and Concentration Distributions: The velocity field of the particle-laden flow has a strong effect on the characteristics of the produced powder. Due to the assumption of a negligible effect of the particles on the aerodynamic behavior of the gas phase, the velocity field establishes the particle residence time which determines the final particle size; shorter residence times lead to smaller particle sizes. Figure 9.3 shows the contours of the magnitude of the velocity field along the B-B' plane (Figure 9.1) [Perez-Fontes and Sohn, 2011].

The gray scale indicates the magnitude of the velocity field. After the funnel, the velocity magnitude shows its highest value of 16 m s^{-1} due to the reduced gas passage area.

Figure 9.4 shows the temperature contours of the mixture inside the reactor along the B-B' plane. The carrier gas and precursor powders enter the reactor at room temperature. Due to the radiation heat transfer from the reactor wall, the temperature of the feeding tubes increases rapidly. The heat transfer from the feeding tubes to the carrier gas and precursor powders is then enhanced.

As a result, the precursors are vaporized and the gaseous mixture is uniformly heated to the maximum reaction temperature by the end of the feeding tubes. After the funnel position, the mixture temperature gradually decreases towards the end of the reactor.

Figure 9.5 shows the mole fraction distributions of the $AlCl_3$ and Mg vapors inside the reactor. These species are fully reacted within a short distance from the end of the funnel, producing aluminum vapor (Figure 9.6). Subsequently, the concentration of aluminum vapor is depleted by nucleation and vapor condensation onto the produced particles. It is noted that no aluminum vapor remains at the reactor outlet.

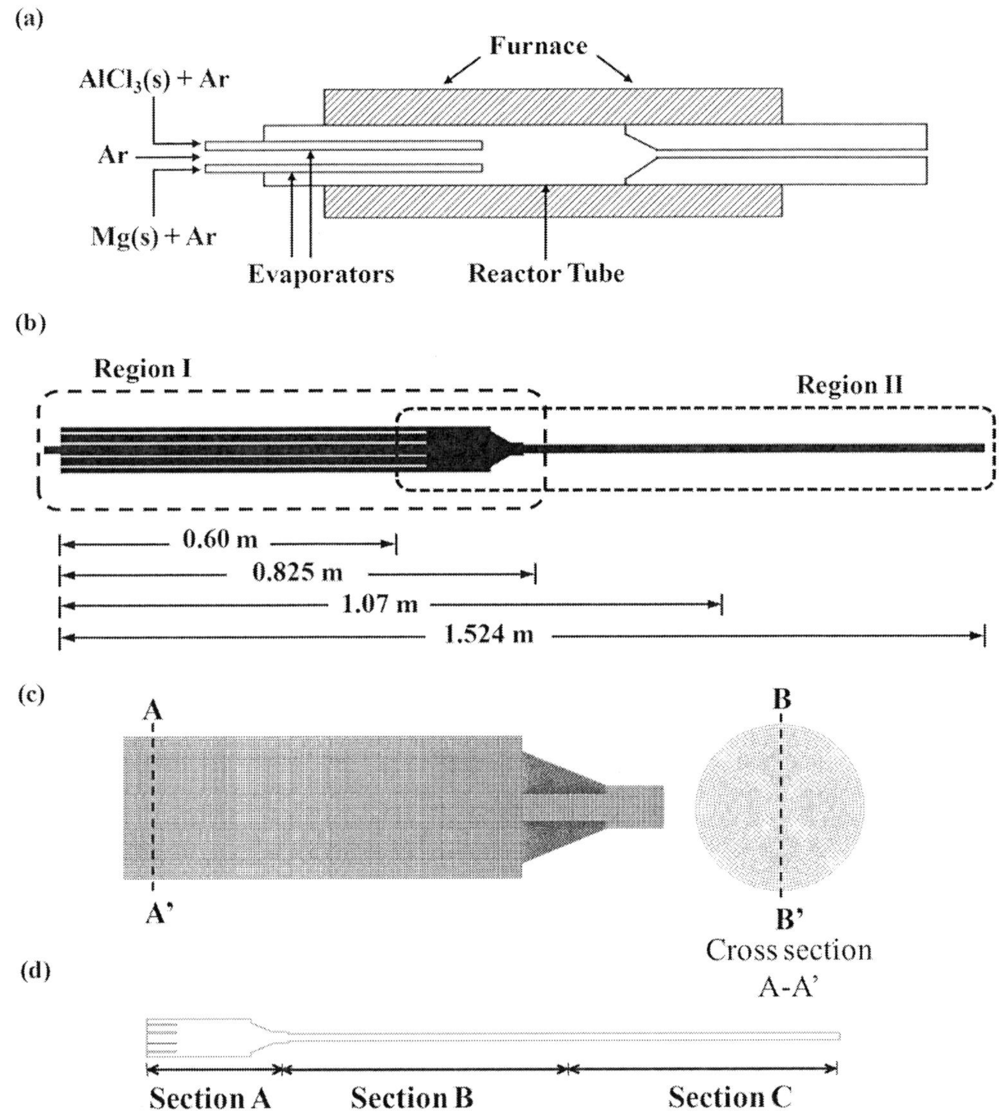

Figure 9.1. Geometry of the simulated reactor: the left side of the figures represents the inlet of the reactants while the right side represents the reactor outlet. (a) Actual reactor, (b) geometries used for the simulation, (c) magnified view of the mesh design, (d) sections for illustration of Region II. [Adapted from Perez-Fontes and Sohn (2011)].

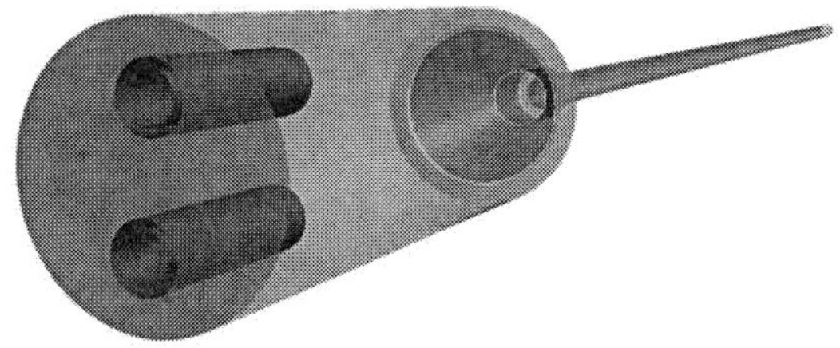

Figure 9.2. Isometric view of the computational geometry. [Adapted from Perez-Fontes and Sohn (2011)].

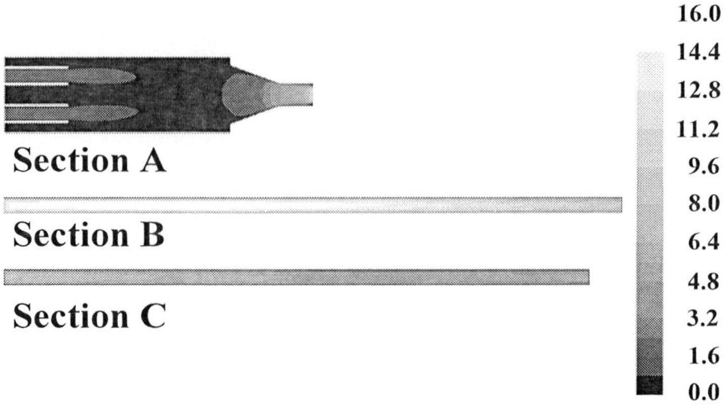

Figure 9.3. Contours of velocity magnitude along the B-B' plane. Scales values in m s^{-1}. [Adapted from Perez-Fontes and Sohn (2011)].

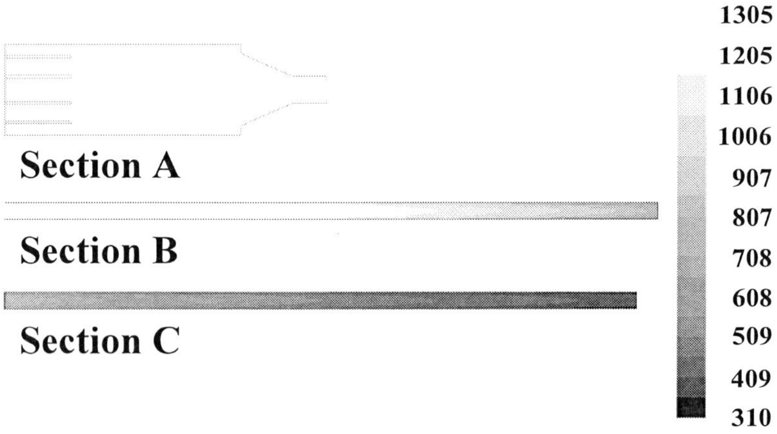

Figure 9.4. Contours of mixture temperature along the B-B' plane. Scale values in K. [Adapted from Perez-Fontes and Sohn (2011)].

Average Particle Size (APS): Figure 9.7 shows the comparison of the simulation results with experimental average particle size as a function of total argon flow rate [Perez-Fontes and Sohn, 2011]. The model predictions showed a reasonable agreement with the experimental data within 10 nm error range without the need of adjustable parameters.

Figure 9.8 shows the contours of average particle size (APS) [Perez-Fontes and Sohn, 2011]. Most nucleation takes place at the entrance of the funnel. Thereafter the particles grow by condensation of aluminum vapor onto the existing particles and coagulation. Since the reaction temperature was higher than the melting point of aluminum, the coalescence rate was extremely fast and thus spherical particles were formed after coagulation. Therefore, the coagulation process determined the final particle size. When the calculations were made without including the coagulation process, the resulting particle sizes were smaller than 10 nm for all experimental conditions, which is much smaller than the experimental values. This indicates that the inclusion of the coagulation phenomenon is critical in correctly predicting the particle size.

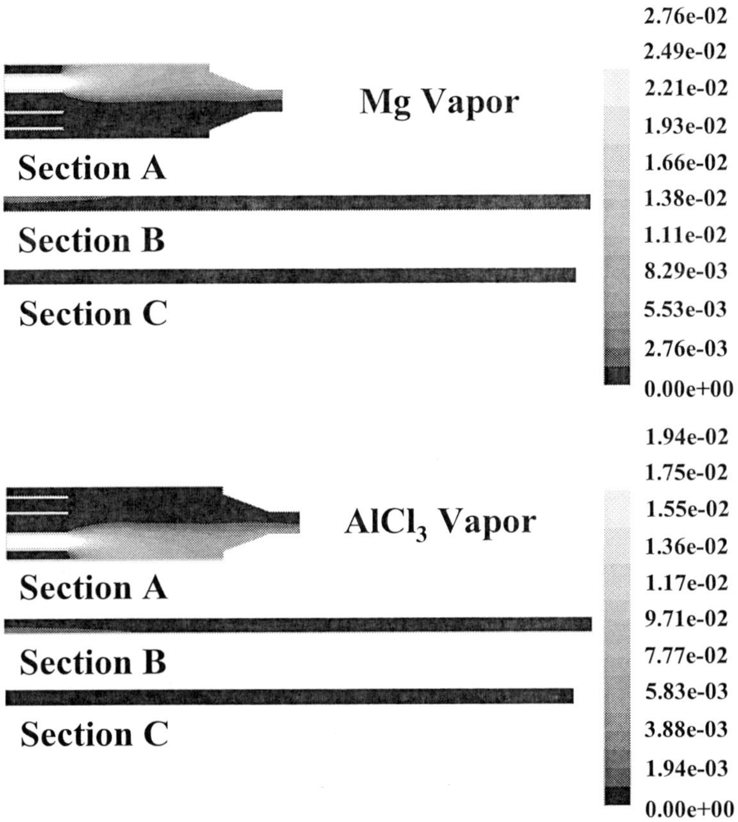

Figure 9.5. Contours of mole fraction of vapor precursors along the B-B' plane. [Adapted from Perez-Fontes and Sohn (2011)].

The distribution of the APS at the outlet plane is given in Figure 9.9. Near the wall, the average particle size is large because the velocity is lower there, which promotes particle growth. However, most of the particles at the outlet were less than 50 nm.

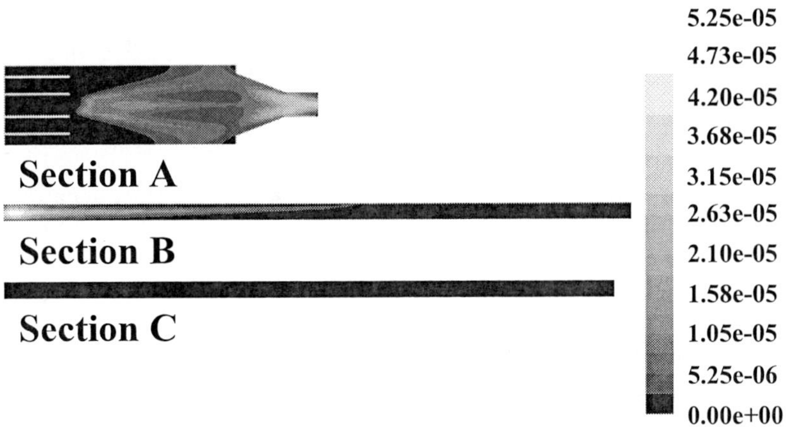

Figure 9.6. Contours of mole fraction of Al vapor along the B-B' plane. [Adapted from Perez-Fontes and Sohn (2011)].

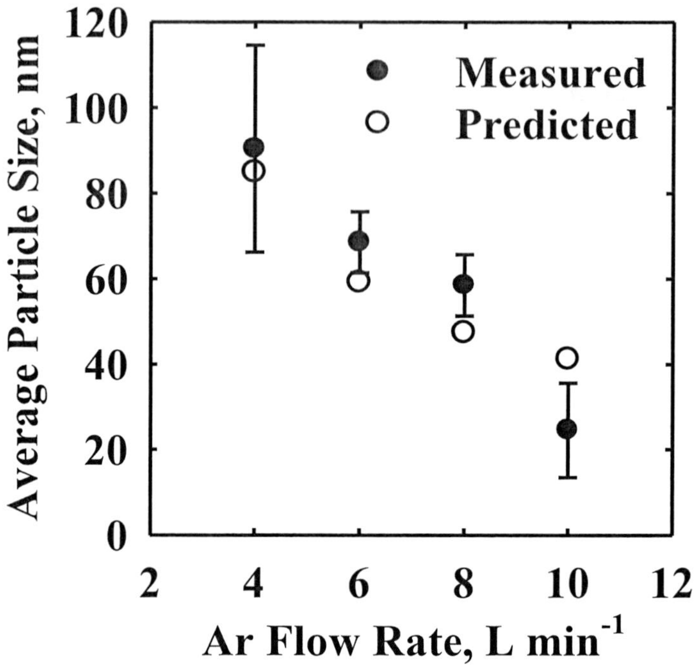

Figure 9.7. Comparison of computed and experimental average particle sizes. [Adapted from Perez-Fontes and Sohn (2011)].

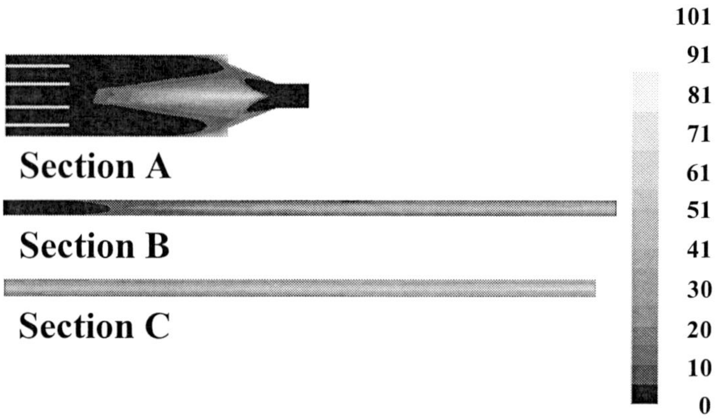

Figure 9.8. Contours of average particle size along the B-B' plane. Scale values in nm. [Adapted from Perez-Fontes and Sohn (2011)].

The geometric standard deviation, σ_{gs}, is commonly used as a measure of the degree of spread of the particle size distribution (PSD) of powders synthesized by vapor-phase processes, even in cases in which the PSD is not truly lognormal. This approach is justified by the fact that several types of aerosol processes are approximately represented by the lognormal distribution [Kodas and Hampden-Smith, 1999]. The general moment expression of the size distribution function for a lognormal distribution is given by

$$\ln \frac{M_k}{M_0} = k \ln d_{pg} + \frac{k^2}{2} \ln^2 \sigma_{gs} \tag{9-26}$$

where d_{pg} is the geometric mean diameter.

By defining equations for the first three moments (k = 0, 1, 2), an expression for the geometric standard deviation can be obtained as follows [Friedlander, 2000]:

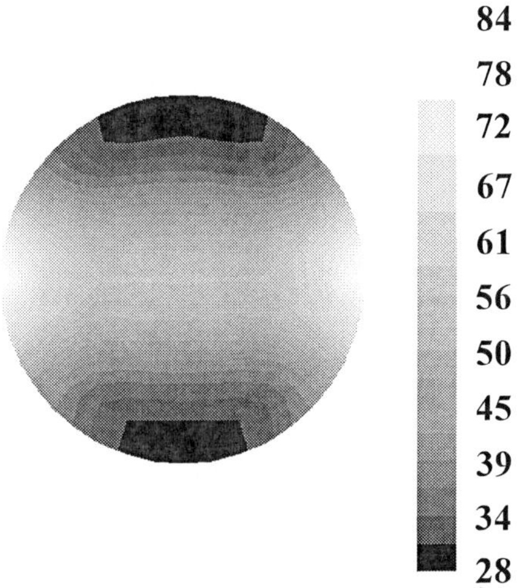

Figure 9.9. Contours of the average particle size at the reactor outlet (A-A' plane). Scale values in nm. [Adapted from Perez-Fontes and Sohn (2011)].

$$\ln^2 \sigma_{gs} = \ln \frac{M_0 M_2}{M_1^2} \tag{9-27}$$

Figure 9.10 shows the effect of the carrier gas flow rate on the profiles of σ_{gs} along the reactor axis. It is seen that all profiles approached a single value of σ_{gs} ($\approx$ 1.44). This value indicates the narrowest particle size distribution that can be obtained at the given conditions. It also indicates that the bulk flow has no effect on the spread of the distribution provided that enough reactor length is available. An interesting piece of information that can be obtained from the σ_{gs} axial profile is the necessary reactor length to attain the narrowest particle size distribution. Thus, reactor overdesign could be avoided.

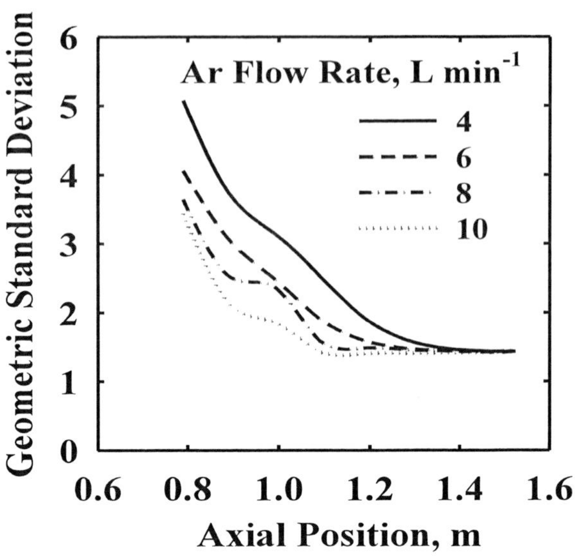

Figure 9.10. Axial profiles of the geometric standard deviation (σ_{gs}). [Adapted from Perez-Fontes and Sohn (2011)].

To summarize, a 3-D computational model incorporating fluid flow, heat and mass transfer, chemical kinetics and population balance was developed for the chemical vapor synthesis of aluminum nanopowder. The nucleation rate was computed using an expression from the classical nucleation theory. The growth rate was obtained by the combined effect of vapor condensation and coagulation. The comparison of the model predictions with the available experimental data showed good agreement under different operating conditions. Computed results indicated that particle coagulation determined the final particle size in the simulated CVS reactor. The model described here can be applied to other similar systems with confidence even without the need of any experimental data, and can be used for scale-up of the process.

9.4. Simulation of SiO_2 Nanopowder Flame Synthesis

Sohn and coworkers [Sohn et al., 2011; Olivas-Martinez et al., 2008; Ji et al., 2007] performed modeling work related to the production of silica nanopowder from waste silicon sludge using a flame spray pyrolysis (FSP) synthesis reactor developed at the Korea Institute of Geoscience and Mineral Resources (KIGAM) [Jang et al., 2007]. The details of the process are described in Chapter 7. Silica nanopowder was synthesized from the thermal oxidation of the volatile precursors: tetraethylorthosilicate (TEOS) and tetramethylorthosilicate (TMOS). Experiments were conducted at three different size scales, laboratory-, bench-, and pilot-scale flame reactors. The present model was used to simulate these three reactors [Sohn et al., 2011; Olivas-Martinez et al., 2008; Ji et al., 2007]. In this chapter, computed results for the simulation of the bench-scale FSP reactor [Sohn et al., 2011] are presented. Figure 9.11 shows a schematic representation of the simulated flame spray pyrolysis (FSP) reactor [Sohn et al., 2011]. It consists of an external-mixing two-fluid nozzle, a burner, a flame cover and a particle collector. The FSP process starts with the formation of a hydrogen-oxygen diffusion flame. A precursor solution containing TEOS and ethanol is then sprayed into the flame through the nozzle. The spray droplets are vaporized and the resulting vapor is reacted producing condensable products that form particles.

The flame synthesis of silica nanopowder by the vapor-phase thermal oxidation of tetraethylorthosilicate (TEOS) is represented by

$$Si(OC_2H_5)_4(g) \rightarrow SiO_2(s) + 4C_2H_4(g) + 2H_2O(g) \quad (9\text{-}28)$$

and

$$Si(OC_2H_5)_4(g) + 12O_2(g) \rightarrow SiO_2(s) + 8CO_2(g) + 10H_2O(g). \quad (9\text{-}29)$$

Equations (9-28) and (9-29) correspond to the thermal decomposition and oxidation of TEOS, respectively.

Simulation Strategy: The main model presented in Section 9.2 was used as a framework to construct a simulation model for this system. It incorporated the presence of the precursor liquid droplets (spray), the combustion process (flame), and the kinetics of nucleation and growth of silica nanoparticles in a flame spray pyrolysis (FSP) reactor [Sohn et al., 2011].

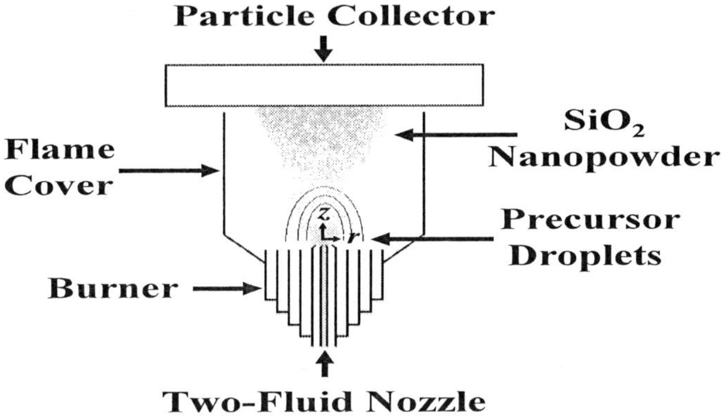

Figure 9.11. Schematic representation of the simulated FSP reactor. [Adapted from Sohn et al. (2011)].

The modeling of the spray liquid droplets was conducted from the Lagrangian viewpoint in which the droplet trajectories are tracked from their injection points. The combustion rate of the fuel was calculated using the eddy-dissipation model [Magnussen and Hjertager, 1976]. The nucleation and growth of silica particles were represented by chemical reaction and Brownian coagulation, respectively. The simulation was conducted in a two-dimensional axisymmetric framework.

Spray Model: The Lagrangian model consists of a coupled set of ordinary differential equations (ODE's) for the velocity, temperature and composition of a droplet along its trajectory with its residence time as the independent variable. In addition, the exchange of momentum, energy and mass between the droplet and the gas phase is computed. The resulting source terms are then added to the gas-phase equations. The model formulation includes the following assumptions: (a) spherical droplets, (b) uniform initial droplet size, and (c) no droplet-droplet interaction.

The equation of motion for a droplet in the Lagrangian framework is

$$\frac{d\vec{u}}{dt} = \frac{3}{4} C_D \frac{\rho_g}{\rho_d} \frac{1}{d_d} \left|\vec{v}_m - \vec{u}\right| \left(\vec{v}_m - \vec{u}\right) + \frac{\left(\rho_d - \rho_m\right)}{\rho_d} \vec{g} \tag{9-32}$$

where $\vec{u}$ is the droplet velocity vector. Equation (9-32) states that the rate of change of droplet momentum is equal to the sum of the forces acting upon it. The first and second terms on the right-hand side represent the aerodynamic drag force and the gravitational acceleration, respectively.

The equation of energy for a droplet moving along its trajectory is

$$m_d c_{p,d} \frac{dT_d}{dt} = h A_d \left(T - T_d\right) + \sum_i \frac{dm_{i,d}}{dt} h_{fg,i} \tag{9-33}$$

where T_d is the droplet temperature. The first term on the right-hand side represent the droplet heating by forced convection. The summation term on the right-hand side of Equation (9-33) accounts for the rate of heat consumed during the vaporization of the species constituting the droplet.

The vaporization of a droplet is driven by the convective mass transfer from the droplet surface to the gas phase and its rate expression is given by

$$\frac{dm_{i,d}}{dt} = A_d M_i k_{c,i} \left(C_{i,s} - C_{i,b}\right). \tag{9-34}$$

The gas-phase concentration at the droplet surface, $C_{i,s}$, is calculated according to Raoult's law. The mass of the droplet, m_d, is then calculated as the sum of the masses of its constituting species:

$$m_d = \sum_i m_{i,d} \tag{9-35}$$

and the density of the droplet ρ_d is computed as the volume-averaged value of the densities of its components:

$$\rho_d = \left(\sum_i \frac{m_{i,d}}{m_d \rho_{i,d}} \right)^{-1}. \tag{9-36}$$

The heat and mass transfer correlations needed to calculate the convective heat and mass transfer coefficients h and $k_{c,i}$, respectively, can be found elsewhere [e.g. Szekely et al., 1976].

Kinetics of Particle Formation and Growth: The Kelvin equation predicts that the critical nucleus size of silica is smaller than the size of its monomer. As a result, the nucleation rate was computed by the rates of the chemical reactions represented by Equations (9-28) and (9-29).

The nucleation rate based on the thermal decomposition of TEOS (Equation (9-28)) is then given by

$$J = k_d C_{\text{TEOS}} N \tag{9-30}$$

with the decomposition rate constant k_d expressed by the following equation [Chu et al., 1991]:

$$k_d = 7.4 \times 10^{10} \exp\left(-\frac{49500}{RT} \right) \tag{9-31}$$

where C_{TEOS} is the molar concentration of TEOS in the vapor phase, and R = 1.987 cal mol^{-1} K^{-1}. The nucleation rate based on the oxidation of TEOS, Equation (9-29), was computed using the eddy-dissipation model [Magnussen and Hjertager, 1976] in which the rate of reaction is limited by the turbulent mixing of the reactants, considering the very rapid chemical kinetics at the high temperatures of the system.

In the flame synthesis process, the growth of silica nanoparticles takes place mostly by Brownian coagulation. In solving the population balance model (Equation (9-8)), a value for the collision efficiency factor used in the expression for the collision frequency function (Equation (9-19)) must be provided. A collision efficiency factor of unity led to particle sizes ten times larger than the experimental value. This result is related to the assumption of infinitely fast coalescence rate. The morphology of silica nanoparticles produced in flames is seldom spherical. Rather, aggregates of primary particles are typically formed. Therefore, the rate of coalescence (sintering) of silica nanoparticles is finite. Various efforts [Lee et al., 2001; Ehrman, 1999; Lehtinen et al., 1996] have been made to develop a model to describe the sintering of silica nanoparticles. However, these models have a semi-empirical nature and their ranges of applicability are limited to the corresponding experimental conditions [Ehrman et al., 1998]. The finite sintering rate of silica nanoparticles is accounted for by setting a collision efficiency factor α smaller than unity. Its value is adjusted to match the experimentally measured average particle size (APS).

Simulation Details: Figure 9.12 shows a schematic representation of the reactor geometry and inlet gas velocities used in the simulation runs [Sohn, 2010]. The inner/outer diameters of the four concentric tubes composing the burner were 32/35, 38/42, 45/48.5, 51.5/55 mm. The precursor solution feed rate was varied from 18.3 to 35.8 mL/min. Table 9-1 shows the experimental conditions used during the simulation runs [Sohn, 2010].

Precursor solution droplets of an initial uniform size of 10 μm were injected into the reactor at the position of the nozzle tip. Five droplet streams were defined at the injection port. Each stream supplied one-fifth of the total mass flow rate of the liquid solution.

The linear velocity of the droplets was computed from the volumetric flow rate of the precursor solution and the cross-sectional area of the nozzle. The concentration of the droplets was defined according to the experimental conditions.

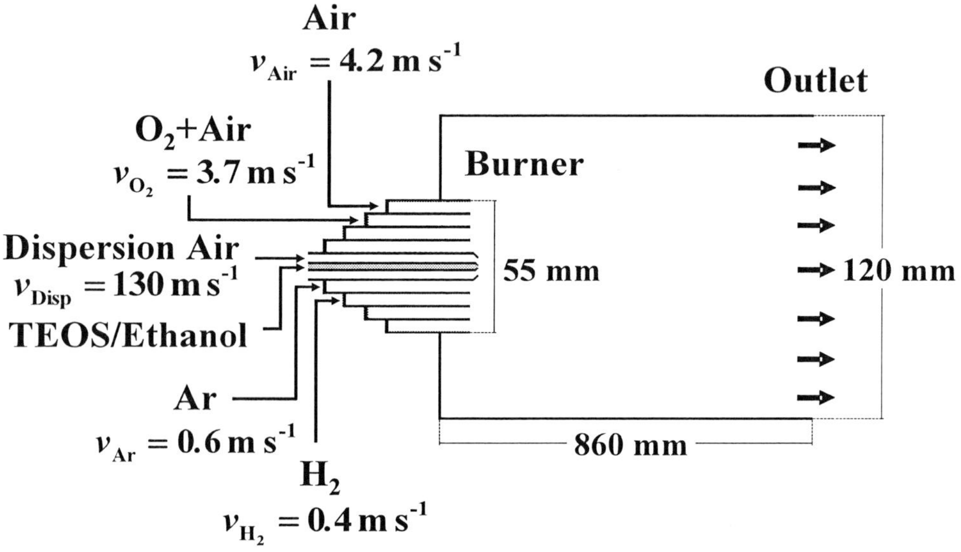

Figure 9.12. Schematic representation of the reactor geometry used in the simulation of the bench-scale FSP reactor. Inlet velocities converted from the experimental flow rates. [Adapted from Sohn (2010)].

Velocity, Temperature and Species Distributions: Figure 9.13 shows the contours of the velocity magnitude inside the FSP reactor [Olivas-Martinez et al., 2008]. The shape of the high velocity contours resembles the jet of a high velocity spray. Figure 9.14 shows the comparison between the observed flame and computed gas temperature contours inside the FSP reactor [Olivas-Martinez et al., 2008].

It is seen that the computer model reproduces reasonably the shape of the flame. More importantly, the computed maximum temperature is close to the calculated adiabatic flame temperature for a hydrogen/oxygen diffusion flame, but somewhat lower than the theoretical value as expected when the components concentrations are decreased by mixing.

Figure 9.15 shows the effect of the dispersion air feed rate on the gas temperature [Sohn et al., 2011]. It is noted that the flame length decreases as the dispersion air feed rate increases. This means that gas-phase species mixing is enhanced and as a result, the rate of the chemical reactions is increased.

Therefore, the fuel species are consumed faster at higher dispersion air feed rates, producing a shorter flame. Figure 9.16 shows that the dispersion air variation dilutes the content of $C_2H_5OH(g)$ in the reactor and that the $C_2H_5OH(g)$ is completely combusted within a small area close to the vicinity of the burner [Sohn, 2010]. The latter result explains why the maximum flame temperature does not vary as the dispersion air feed rate is varied. Figure 9.17 shows that the concentration of TEOS(g) also decreases as the dispersion air increases.

By comparing the concentration contours shown in Figures 9-16 and 9-17, it can be seen that the region of the highest ethanol concentration is located closer to the nozzle than that of the highest TEOS. This means the ethanol vaporizes faster than TEOS. This result is in agreement with the boiling points of ethanol and TEOS which are 78.4 and 169°C, respectively.

Table 9.1. Experimental Conditions for the Bench-Scale FSP Reactor [Sohn, 2010]

Experimental Condition	Flow Rate of Dispersion Air	Feed Rate of Precursor	Concentration of Precursor	Particle Size
	L/min	mL/min	% vol.	Nm
Variation of the Dispersion Air Feed Rate				
JH060712_03	6	26.2	30	20.54
JH060714_03	8.3	26.2	30	13.41
JH060713_01	11	26.2	30	11.42
Variation of the Precursor Solution Feed Rate				
JH060719_02	11	18.3	50	10.57
JH060719_01	11	26.2	50	11
JH060802_01	11	35.8	50	14.41

Figure 9.18 shows that the flame length is increased as the precursor feed rate increases. This is caused due to the addition of more fuel (ethanol) to the reactor. Here, the dispersion air feed rate is maintained constant, thus the increase in the rate of combustion is not related to mixing but to an increase in the fuel concentration as shown in Figure 9.16.

Figure 9.13. Contours of gas velocity magnitude inside the bench-scale FSP reactor. Scale values in m s^{-1}. [Adapted from Olivas-Martinez et al. (2008)].

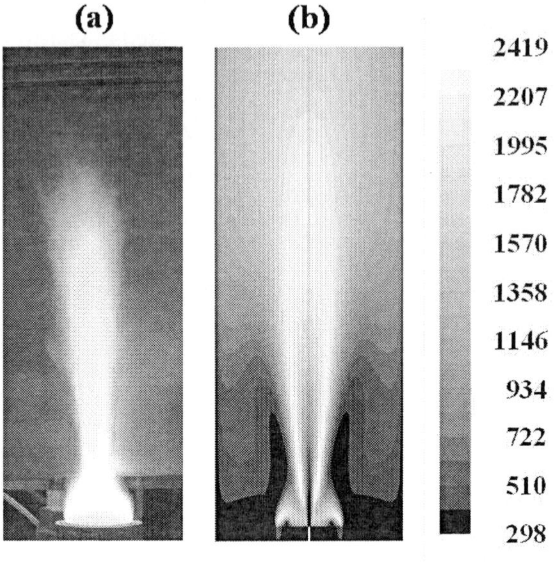

Figure 9.14. Photograph of the flame (a) and temperature contours (b) for the bench-scale flame reactor. Scale values in K. [Adapted from Sohn (2010)].

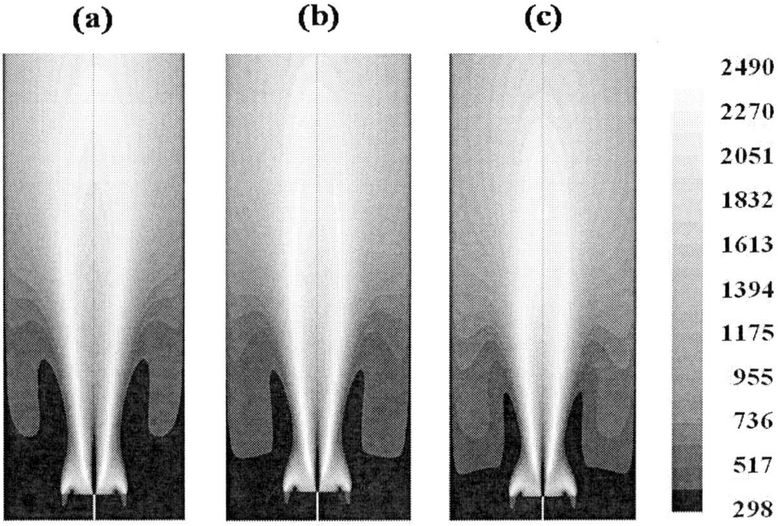

Figure 9.15. Contours of gas temperature in the entire simulated reactor geometry. Dispersion air flow rates: (a) 6, (b) 8.3, and (c) 11 L/min. Values in K. [Adapted from Sohn (2010)].

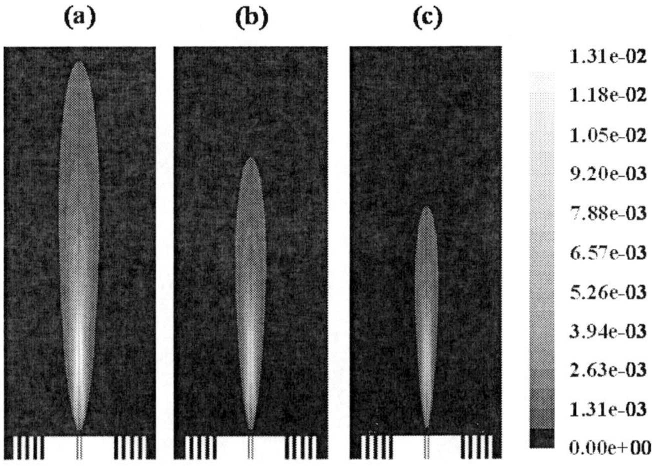

Figure 9.16. Contours of molar concentration of $C_2H_5OH(g)$ in the vicinity of the nozzle tip. Dispersion air flow rates: (a) 6, (b) 8.3, and (c) 11 L/min. Values in mol L^{-1}.The contours area shown is 0.029 x 0.15 m^2. [Adapted from Sohn (2010)].

Effect of Operating Conditions on the Particle Size Distribution: The particle size distribution (PSD) of silica nanopowder was computed by solving the population balance model (PBM) considering particle nucleation and coagulation [Sohn et al., 2011]. The nucleation rate was computed with no adjustable parameters and only the collision efficiency factor α was adjusted to match the experimentally measured average particle size. A value of $\alpha = 4\times10^{-4}$ was found to provide reasonable results for all experimental conditions. As discussed above, the value of the collision efficiency factor α represents the coalescence of primary particles. Such a small value can thus be expected since the coalescence rate is not only finite but also slower than the rate of particle coagulation. The comparison between the experimental and predicted number average particle diameters was done by computing the cup-mixing average diameter from the obtained results.

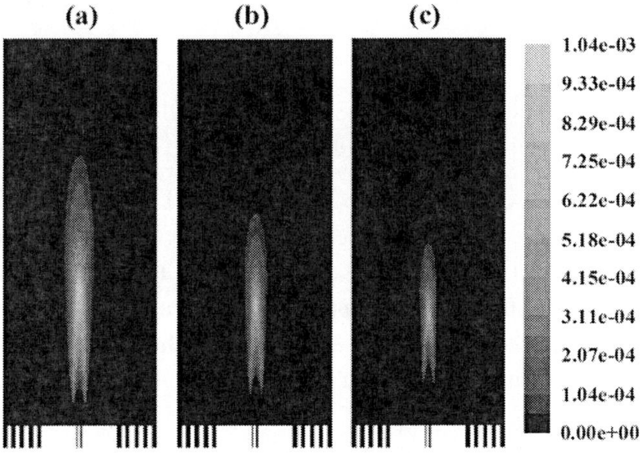

Figure 9.17. Contours of molar concentration of TEOS(g) in the vicinity of the nozzle tip. Dispersion air flow rates: (a) 6, (b) 8.3, and (c) 11 L/min. Values in mol L^{-1}.The contours area shown is 0.029 x 0.15 m^2. [Adapted from Sohn (2010)].

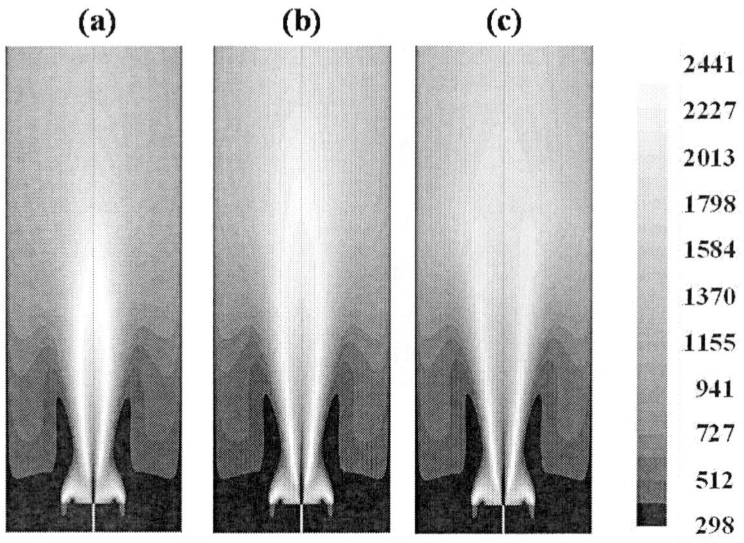

Figure 9.18. Contours of gas temperature in the entire simulated reactor geometry. Precursor solution feed rates: (a) 18.3, (b) 26.2, and (c) 35.8 mL/min. Values in K. [Adapted from Sohn (2010)].

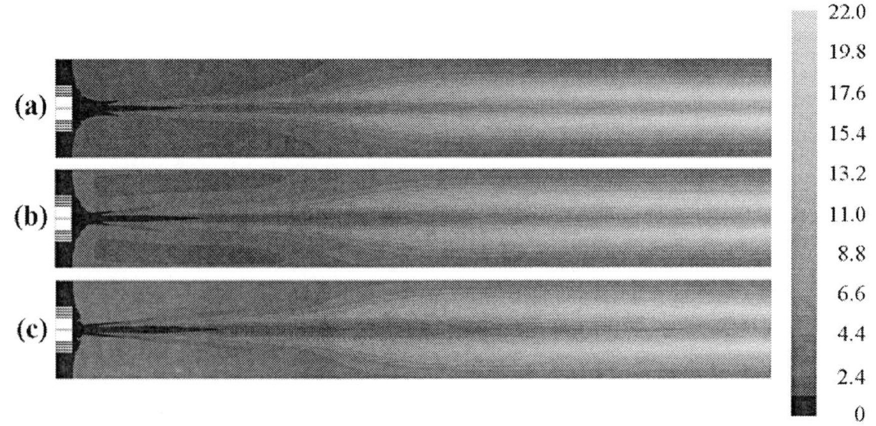

Figure 9.19. Contours of average particle diameter inside the FSP reactor. Precursor solution feed rate: (a) 18.3, (b) 26.2, and (c) 35.8 mL/min. Scale values in nm. [Adapted from Sohn et al. (2011)].

Figure 9.19 depicts the contours of the average particle diameter inside the reactor [Sohn et al., 2011]. It is of interest to note that the particle diameter is larger close to the center line of the reactor and smaller close to the wall. Particles nucleate in the flame and are carried by the bulk flow. This causes the particle concentration to be higher near the reactor axis than close to the walls. The coagulation rate in the center of the reactor is therefore enhanced by the high particle concentrations and high temperature. As a result, particle size is larger around the reactor axis.

The comparison between the experimental and computed average particle diameters as a function of the dispersion air feed rate is presented in Figure 9.20 [Sohn et al., 2011]. The model predictions showed reasonable agreement with the experimental data. As the dispersion air feed rate increases the average particle diameter decreases.

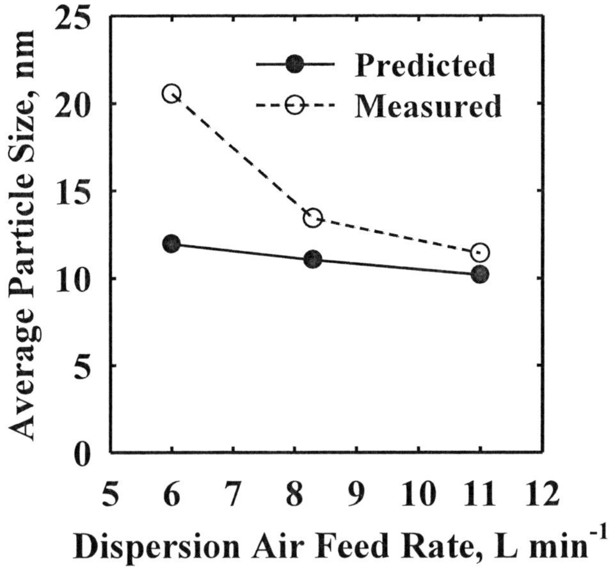

Figure 9.20. Effect of the dispersion air feed rate on the average particle diameter. Precursor liquid solution with feed rate of 26.2 mL/min and composition of 30 % TEOS by volume. [Adapted from Sohn et al. (2011)].

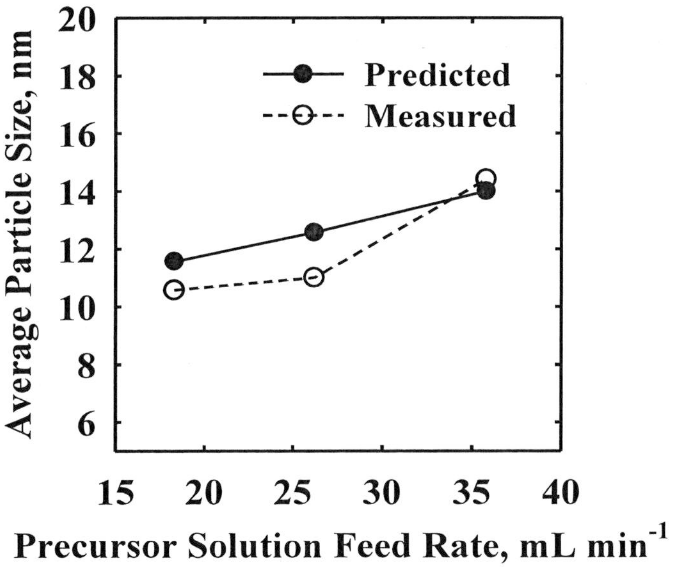

Figure 9.21. Effect of the precursor solution feed rate on the average particle diameter. Flow rate of dispersion air of 11 L/min and a precursor liquid solution with a composition of 50 % TEOS by volume. [Adapted from Sohn et al. (2011)].

The decrease in the concentration of TEOS(g) and the shortening of the flame length as the dispersion air increases are likely the dominant factors. Considering the fact that the particle nucleation is controlled by gas-phase reactions of TEOS(g) and particle growth by Brownian coagulation, the decrease of the precursor concentration with increasing dispersion air flow rate will limit the maximum attainable particle diameter. In addition, the flame length plays an important role on the particle formation and growth by affecting the length of time reactants and particles remain in the reaction zone.

Figure 9.21 shows the comparison between the experimental and computed average particle diameters when the precursor feed rate was varied [Sohn et al., 2011]. The model is capable of reproducing reasonably well the magnitude and trend observed for the average particle diameter as the precursor feed rate increases.

In summary, this section presented CFD modeling results for the simulation of a flame spray pyrolysis (FSP) process for the synthesis of silica nanopowder from a volatile precursor (tetraethylorthosilicate (TEOS)). A Lagrangian model was incorporated to represent the motion and the rate of vaporization of the spray droplets.

The nucleation rate was computed based on the rates of thermal decomposition and oxidation of the volatile precursor. Brownian coagulation was used as the particle growth mechanism. Computed results showed that the model is capable of reproducing the experimental data obtained under different conditions using a single value of the collision efficiency factor. Overall, the model has proved to be a promising a tool for the design and scale-up of industrial-size FSP units.

9.5. SIMULATION OF WC NANOPOWDER CVS

As detailed in Chapters 5 and 6, tungsten carbide (WC) is widely used as a hard material in many industrial applications such as metalworking, drilling and mining. It has been shown that the reduction of WC grain size provides a significant improvement in the mechanical properties, lowers the sintering temperatures, and shortens the compaction time needed to attain dense but small grain structures [Fu et al., 2001; Zawrah, 2007].

Although considerable work to produce nanosized WC powder by gas-phase reactions has been conducted, the mechanisms of WC particle formation and growth are still not well understood. In this section, simulation work [Ryu et al., 2010c] using the modeling approach described in Section 9.2 to describe a chemical vapor synthesis (CVS) reactor for the preparation of WC nanopowder from tungsten hexachloride will be presented. WCl_6 was used as the precursor with reactant gases H_2 and CH_4 as the reducing and carburizing agents, respectively. The model was validated with experimental data in terms of average particle size and concentration of unreacted precursor at the outlet. The contours of temperature, velocity, species concentration and particle size distribution (PSD) inside the tubular reactor were computed.

Total six species considered in this simulation were WCl_6, CH_4, H_2, WC, HCl, and Ar based on the following overall chemical reaction:

$$WCl_6(g) + CH_4(g) + H_2(g) = WC(s) + 6HCl(g) \qquad (9\text{-}37)$$

Simulation Strategy: The terms for this system that need to be incorporated into the CFD modeling equations described in Section 9.2 are developed below.

Kinetics of Particle Formation and Growth: The mechanisms of nucleation and growth of WC nanoparticles from the simultaneous reduction and carburization of WCl_6 are not well understood. The Kelvin equation [Kodas and Hampden-Smith, 1999] predicts that the critical nucleus size of a WC particle is smaller than its monomer size. Hence, the nucleation of WC particles takes place by a reaction-controlled process. Here, the nucleation rate of WC is given by a second-order reaction rate law, which is dependent on the concentrations of WCl_6 and CH_4. The nucleation rate was thus expressed by

$$J = k_n N C_{\mathrm{WCl}_6,b} C_{\mathrm{CH}_4,b} \tag{9-38}$$

where J is the nucleation rate ($\#\cdot s^{-1}\cdot m^{-3}$) and k_n is the nucleation rate constant ($m^3\cdot kmol^{-1}\cdot s^{-1}$).

The growth rate under the control of chemical reaction kinetics [Kodas and Hampden-Smith, 1999; Bolsaitis et al., 1987] is expressed by

$$G_c = k_g C_{\mathrm{WCl}_6,b} C_{\mathrm{CH}_4,b} \tag{9-39}$$

where k_g represents the growth rate constant ($m^7\cdot kmol^{-2}\ s^{-1}\cdot$). It is noted that even though hydrogen is included in the overall reaction (Eq. (9-37)), the proposed rate expressions for particle nucleation (Eq. (9-38)) and growth (Eq. (9-39)) are not dependent on the hydrogen concentration, because hydrogen is in large excess.

The volumetric rate of growth of a particle under the control of mass transfer, assumed to be driven by the limiting reactant (WCl_6), is given by

$$\frac{dv_p}{dt} = V_{\mathrm{WC}} \cdot \pi d_p^2 \cdot k_{c,\mathrm{WCl}_6} \left(C_{\mathrm{WCl}_6,b} - C_{\mathrm{WCl}_6.s} \right) \tag{9-40}$$

where v_p is the particle volume (m^3) and the subscripts b and s represent bulk and surface, respectively. Here, it was assumed that concentration of $WCl_6(g)$ at the particle surface is negligible compared to its concentration at the bulk since Reaction (9-37) has a large equilibrium constant.

The growth rate in terms of the particle diameter is obtained by substituting the expression for the particle volume ($v_p = \pi d_p^3 / 6$) into Eq. (9-40), which yields

$$\frac{dd_p}{dt} = 2V_{\mathrm{WC}} \cdot k_{c,\mathrm{WCl}_6} \cdot C_{\mathrm{WCl}_6,b} \cdot \tag{9-41}$$

The mass transfer coefficient, k_{c,WCl_6}, is calculated based on the assumption that the particles and gaseous mixture move at the same velocity. Thus, the correlation of Ranz and Marshall [Szekely et al., 1976] gives

$$\mathrm{Sh} = \frac{k_{c,\mathrm{WCl}_6} d_p}{D_{\mathrm{WCl}_6 - mix}} = 2.0 \tag{9-42}$$

where Sh is the Sherwood number. By substituting Eq. (9-42) into Eq. (9-41), the particle growth rate under the control of mass transfer, G_m, is given by

$$G_m = 4V_{WC}D_{WCl_6-mix}\frac{M_0}{M_1}C_{WCl_6,b} \quad (9\text{-}43)$$

where M_0 is the zeroth moment (#·m^{-3}) of the particle size distribution (PSD) and M_1 is the first moment (m·m^{-3}) (thus, m_1/m_0 represents the number-weighted average particle size).

The overall growth rate under the combined control of surface chemical reaction and mass transfer can be expressed by

$$G_{overall} = \frac{G_c G_m}{G_c + G_m} \quad (9\text{-}44)$$

In this system studied, the growth rate under the control of surface chemical reaction was the limiting factor, which means that the growth rate under the control of mass transfer is much faster than the growth rate by the surface chemical reaction. This is because gas-solid mass transfer is extremely fast for small particles [Szekely et al. 1976 p. 13].

In terms of the particle growth rate, the rate of increase of the solid mass per unit volume of the system in the overall mass balance is given by

$$Source = (\pi / 2)M_2 G_{overall}\rho_p \quad (9\text{-}45)$$

where ρ_p is the particle density (kg·m^{-3}) of WC(s) and M_2 is the second moment of the PSD (m^2 ·m^{-3}), which represents the total surface area of particles per unit volume of the gaseous mixture.

Simulation Details: Initial 3-D simulation runs considering the entire reactor geometry for reaction temperatures of 1200, 1300 and 1400°C showed that all species were mixed homogeneously before the reactant gases approached a temperature of 600, 720, 835°C, respectively. These values were then adopted as the inlet temperatures. The 3-D simulations also showed that the flow of the gaseous mixture became symmetric around the reactor centerline before the reaction zone was reached. This evidence allowed the authors to do the simulation in a 2-D axisymmetrical framework.

Figure 9.22 shows the experimental wall temperature profile and the simulated reactor geometry. The inner diameter and total length of reactor simulated were 5.4 cm and 38 cm, respectively. The temperature at the inlet was 835°C and after 3.5 cm and 12 cm in distance from the inlet it was increased to 1000°C and 1400°C, respectively. According to Hojo et al. [1978] and experimental work by Ryu et al. [2008c], substantial WC formation from WCl_6 starts to occur above 1000°C. The mass fractions of WCl_6, CH_4, and H_2 at the inlet were 0.0347, 0.0483, and 0.0152, respectively, based on the experimental conditions [Ryu et al., 2008c].

Determination of Nucleation Rate and Growth Rate Constants: The nucleation rate and growth rate are important factors that determine the particle size distribution (PSD). However, it is extremely difficult to determine the nucleation rate constant (k_n) and growth rate constant (k_g) by experimental measurements for the synthesis of WC nanoparticles.

Because there were no measured data on these parameters, these parameters were first obtained by adjusting k_n and k_g and comparing the computed results with experimental measurements of average particle size and the coefficient of variation in some cases. In the experimental results, in which WC was synthesized under the conditions of WCl_6 feeding rate of 0.04 g/min, H_2 flow rate of 0.25 L/min (25°C, 86.1 kPa), CH_4 flow rate of 0.1 L/min (25°C, 86.1 kPa), and Ar flow rate of 0.75 L/min (25°C, 86.1 kPa), and the reaction temperature of 1400°C, the average particle size based on specific surface area measurement was 22.2 nm and the coefficient variation (C.V.) was 0.18.

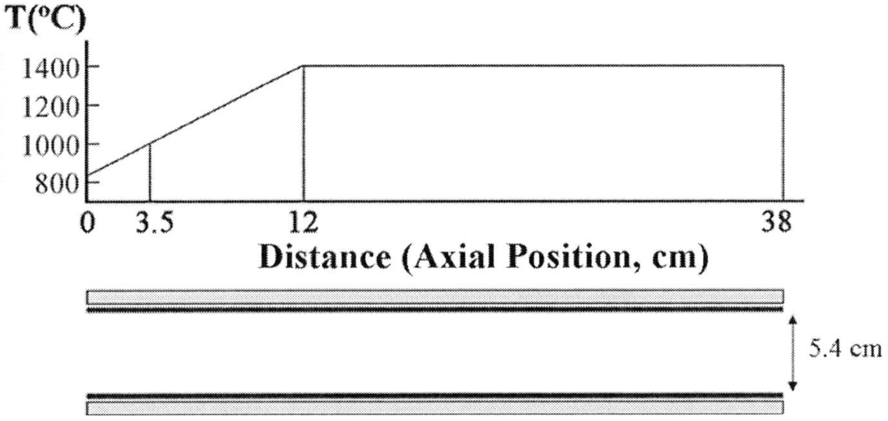

Figure 9.22. Reactor wall temperature profile and geometry. [Adapted from Ryu et al. (2010c)].

The experimental average particle size, $\bar{d}_{p,exp}$, was calculated from the measured specific surface area, $S_{p,exp}$, under the assumption of spherical particles:

$$\bar{d}_{p,exp} = \frac{6}{\rho_p S_{p,exp}} \tag{9-46}$$

The coefficient of variation (C.V.) means the degree of spread of the size distribution. It is represented as the ratio of the standard deviation of a distribution (σ_p) to its mean size (μ_p), as expressed below:

$$\text{C.V.} = \frac{\sigma_p}{\mu_p}. \tag{9-47}$$

In the simulation results, the average particle size based on specific surface area, often called Sauter mean diameter, was derived as follows:

$$d_{p,Sauter} = \frac{6}{\rho_p S_p} = \frac{6}{\rho_p \frac{A_p}{\rho_p V_p}} = \frac{6V_p}{A_p} = \frac{6\frac{\pi}{6}\int_0^\infty d_p^3 n(d_p)dd_p}{\pi \cdot \int_0^\infty d_p^2 n(d_p)dd_p} = \frac{M_3}{M_2} \qquad (9\text{-}48)$$

where $d_{p,Sauter}$ is the computed local average particle size (m).

The local coefficient of variation (c.v.) is calculated in terms of the moments of the PSD as follows [Randolph and Larson, 1988]:

$$\text{c.v.} = \frac{\left(\frac{1}{M_2}\int_0^\infty \left(d_p - d_{p,Suater}\right) d_p^2 n(d_p)dd_p\right)^{1/2}}{d_{p,Sauter}} = \left(\frac{M_4 M_2}{M_3^2} - 1\right)^{1/2} \qquad (9\text{-}49)$$

To compare the experimental and simulation results, the corresponding average particle size of all the particles leaving the reactor was computed taking into consideration the radius-dependent flow rates of particle volume and surface area, as follows:

$$\bar{d}_{p,Sauter} = \frac{\dot{M}_3}{\dot{M}_2} \qquad (9\text{-}50)$$

where

(9-51)

is the flow rate of the k-th moment of the PSD and $\vec{A}$ is the cross-sectional area vector. Here, it was assumed that the particles have the same velocity as the gas. Similarly, the C.V. at the reactor outlet is given by

$$\text{C.V.} = \left(\frac{\dot{M}_4 \dot{M}_2}{\dot{M}_3^2} - 1\right)^{1/2} \qquad (9\text{-}52)$$

However, in the experimental results, the particle size measurement was from the rather limited number of particles. Thus, in view of this, average particle size, complete consumption of WCl_6 at outlet as observed experimentally, and numerical convergence were used as the criteria to determine the rate constants. The values $k_n = 1\times10^{-4}$ $m^3 \cdot mol^{-1} \cdot s^{-1}$ and $k_g = 7.5\times10^{-6}$ $m^7 \cdot mol^{-2}$ s^{-1} were selected since they yielded an average particle size (20.4 nm) which is close to the experimental value (22.2 nm), resulted in no unreacted WCl_6, and allowed numerical solution to converge.

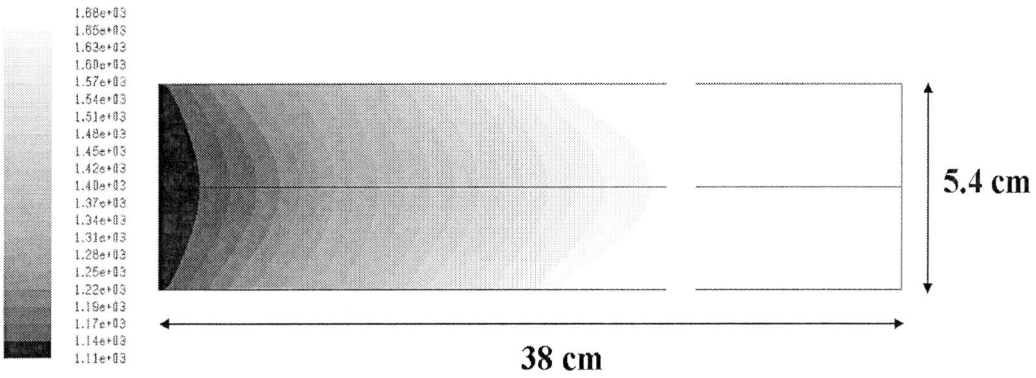

Figure 9.23. Contours of temperature (K) inside the reactor. [Adapted from Ryu et al. (2010c)].

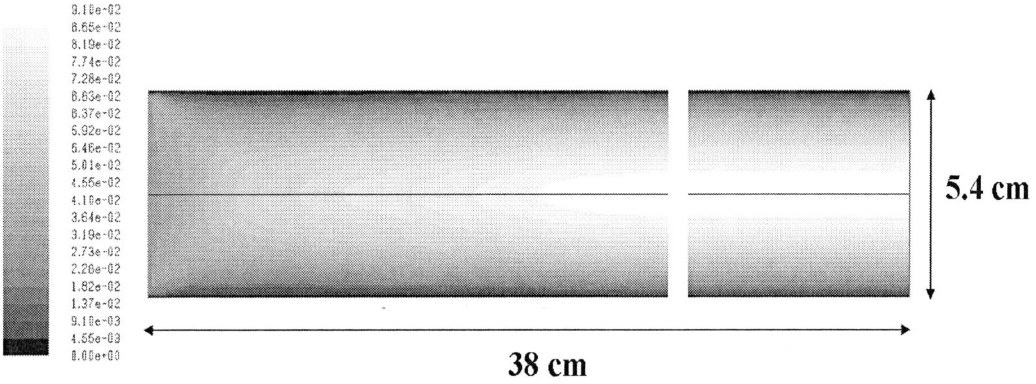

Figure 9.24. Contours of velocity magnitude (m/s). [Adapted from Ryu et al. (2010c)].

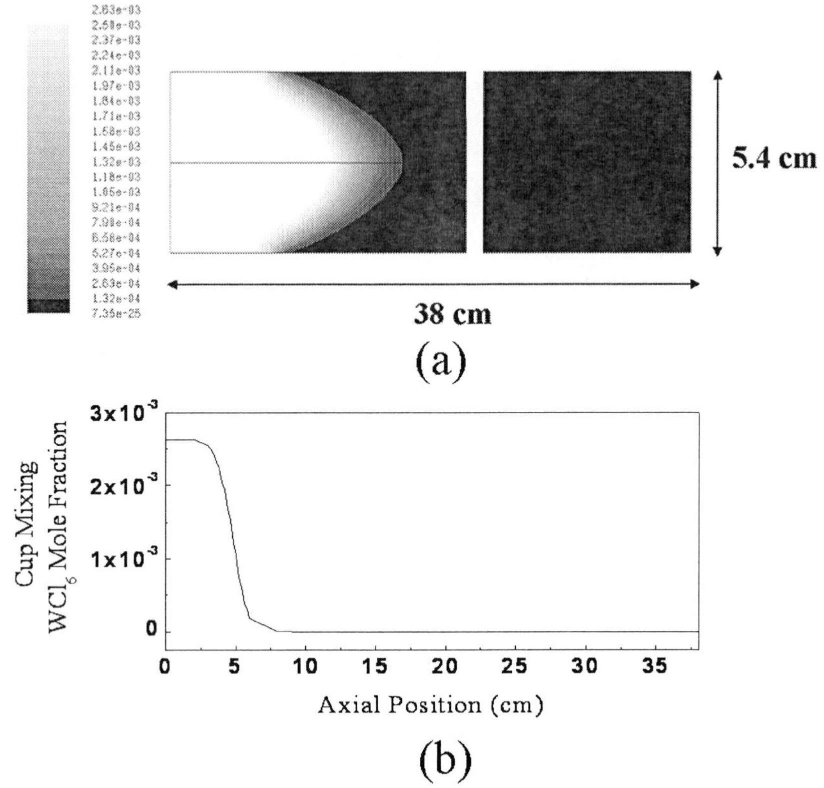

Figure 9.25. Contours (a) and profile (b) of mole fraction of WCl_6. [Profile is based on the local cup-mixing value.] [Adapted from Ryu et al. (2010c)].

Temperature, Velocity and Concentration Profile: From the initial computation, the nucleation rate constant and growth rate constant were fixed at 1×10^{-4} $m^3{\cdot}mol^{-1}{\cdot}s^{-1}$ and 7.5×10^{-6} $m^7{\cdot}mol^{-2}{\cdot}s^{-1}$, respectively, for further simulation work. Figure 9.23 shows the temperature contours with fixed nucleation and growth rate constants obtained numerically as mentioned above.

The velocity magnitude of the mixture is shown in Figure 9.24, in which the total linear velocity of reactant gases at the inlet was 0.030 $m{\cdot}s^{-1}$. The inlet and the plateau wall temperatures were 835°C and 1400°C, respectively. The highest velocity was observed at the center of reaction zone and the lowest velocity was observed near the wall of the reactor. These converged results are reasonable as far as the temperature and velocity profile are concerned.

The contours and profile of mole fraction of WCl_6 in Figure 9.25 show that WCl_6 was consumed completely and that the reaction with hydrogen and methane occurred rapidly. Figures 9-26 and 9-27 show the contours of mole fractions of hydrogen and methane, respectively, in which they are seen to be consumed rapidly for WC formation. These results were reasonable considering that the reaction in the vapor phase occurs rapidly.

Figures 9-28 and 9-29 show the contours and profiles of nucleation rate and growth rate obtained by applying the kinetic constants as discussed previously. These results were in good agreement with the WC formation reaction, in which the nucleation and growth of WC particles start rapidly above 1000°C. It is also clearly seen that the nucleation and growth of WC particles take place rapidly as $WCl_6(g)$ is consumed.

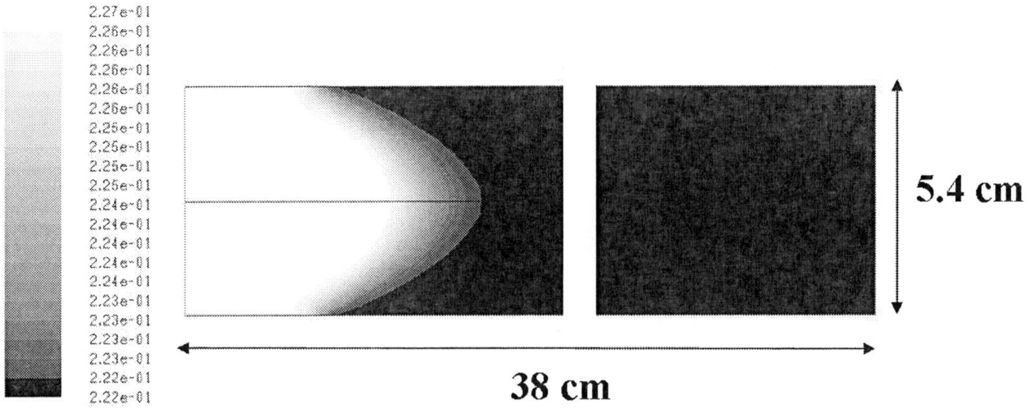

Figure 9.26. Contours of mole fraction of H_2. [Adapted from Ryu et al. (2010c)].

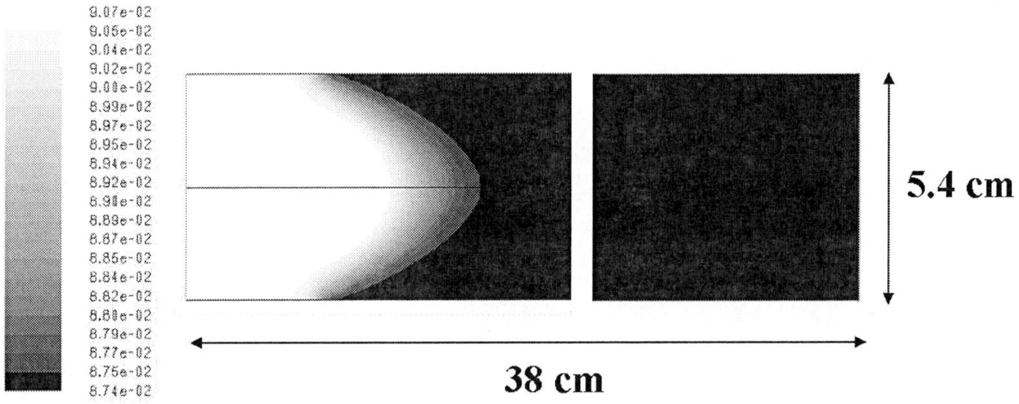

Figure 9.27. Contours of mole fraction of CH_4. [Adapted from Ryu et al. (2010c)].

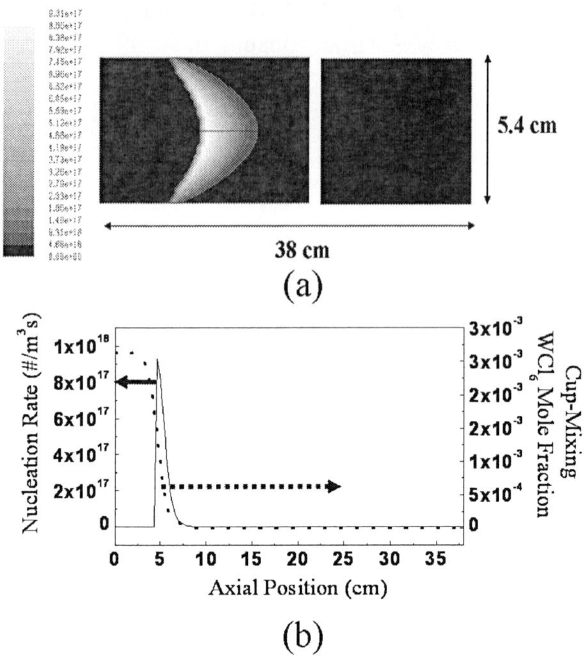

Figure 9.28. Contours (a) and profile (b) of nucleation rate of WC particles ($\#\cdot m^{-3}\cdot s^{-1}$). [Profile is based on the local value along the reactor axis.] [Adapted from Ryu et al. (2010c)].

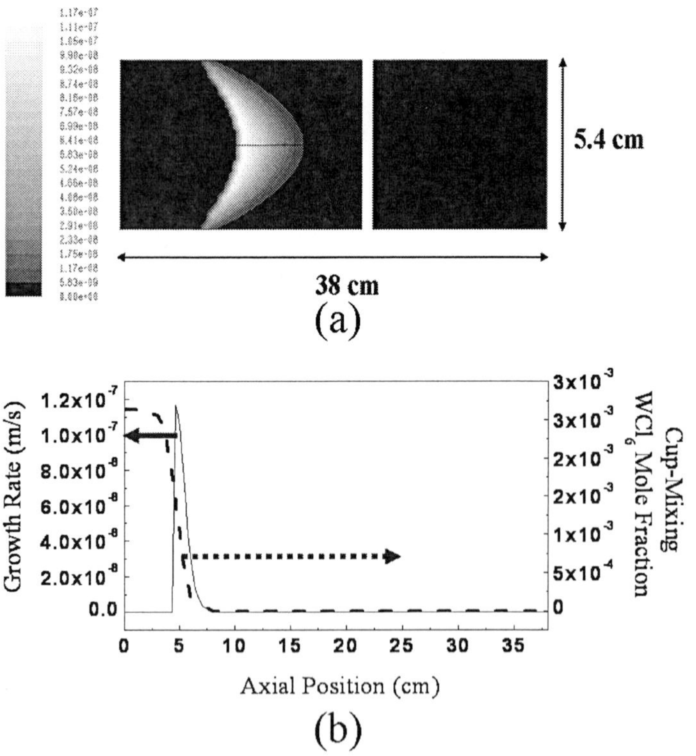

Figure 9.29. Contours (a) and profile (b) of growth rate of WC particles ($m\cdot s^{-1}$). [Profile is based on the local value along the reactor axis.] [Adapted from Ryu et al. (2010c)].

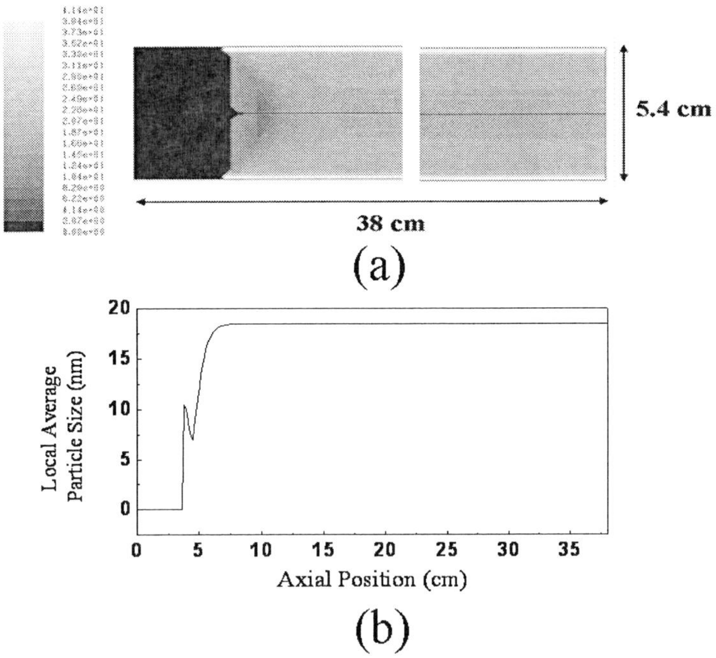

Figure 9.30. Contours (a) and profile (b) of WC average particle size (nm). [Profile is based on the local value along the reactor axis.] [Adapted from Ryu et al. (2010c)].

Since the flow regime in this reactor is laminar, there is a parabolic velocity profile. As a result, the mass transfer by convection in the axial direction is faster in the center of the reactor than near the walls. Therefore, at a fixed axial position the concentration of precursors at the center is higher than that near the wall (as shown in Figures 9-25 – 9-27). Consequently, the rates of nucleation and growth of the particles, which are proportional to the concentration of the gaseous precursors, are higher at the center and lower near the wall, at a fixed axial position.

Particle Size Distribution: As mentioned earlier, the nucleation rate and growth rate constants are two important factors that determine the particle size distribution.

Figure 9.30 shows the contours and profile of computed WC average particle size in the tubular reactor, in which the nucleation rate constant and growth rate constant were fixed at 1×10^{-4} $m^3\cdot mol^{-1}\cdot s^{-1}$ and 7.5×10^{-6} $m^7\cdot mol^{-2}\cdot s^{-1}$, respectively. From the results, it is observed that the largest average particle size of 41 nm was near the reactor wall, where the linear velocity of mixture is lowest, and the average particle size at the outlet was 21 nm, as shown in Figure 9.30(a). Figure 9.30(b) shows the profile of area-weighted average particle size along the axial position. The decrease of its value around 4 cm can be explained by the nucleation rate being a maximum at this point, i.e. the number of particles is maximum.

If more detailed experimental data on the average particle size are available, additional model verification and validation can be conducted. The nucleation rate and growth rate constants (k_n and k_g) obtained at 1400°C of reaction temperature were applied to verify these constants by the comparison of the average particle size obtained at different CH_4/WCl_6 molar ratios in the reaction. Figure 9.31 shows the variation of the average particle size as a function of the CH_4/WCl_6 molar ratio. For all CH_4/WCl_6 ratios, the methane concentration was varied at a fixed WCl_6 concentration and methane was always in excess. The simulation results show that the average particle size first decreased with the increase of the methane concentration up to the CH_4/WCl_6 ratio of 12 where the smallest average particle size was observed. After this ratio, the average particle size increased with an increase in methane concentration. This behavior is the result of the coupled interaction among the rates of nucleation and growth of WC particles as the concentration of methane increased. However, it is important to note that the variation of the average particle size with CH_4/WCl_6 ratio was not large. The experimentally determined average particle size was not affected by methane concentration. Thus, the rather small variation (4 nm) in the simulation can be considered to be consistent with the experimental results.

The experimental results indicate that the average particle size decreased with a decrease in reaction temperature. To determine the best set of nucleation rate and growth rate constants at different reaction temperatures, the same approach of adjusting these values to obtain the average particle size close to the experimental results was applied. At each temperature, a set of constants was obtained to determine the temperature dependence of the nucleation and growth phenomena. The optimum values of nucleation rate and growth rate constants that matched the experimental data at 1300°C and 1200°C were 1×10^{-4} $m^3\cdot mol^{-1}\cdot s^{-1}$ and 6×10^{-6} $m^7\cdot mol^{-2}\cdot s^{-1}$, and 1×10^{-4} $m^3\cdot mol^{-1}\cdot s^{-1}$ and 3.3×10^{-6} $m^7\cdot mol^{-2}\cdot s^{-1}$, respectively.

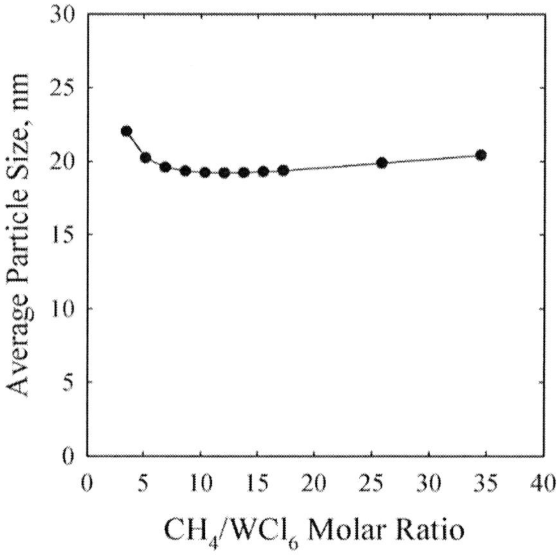

Figure 9.31. Effect of the CH_4/WCl_6 molar ratio on the average particle size. [Adapted from Ryu et al. (2010c)].

These results indicate that the nucleation rate constant did not change with the variation of the reaction temperature. In the experimental work [Ryu et al., 2008c], the reaction temperature is defined as the maximum temperature reached inside the reactor by external heating with a furnace as illustrated in Figure 9.22. Considering the fact that significant WC formation occurs above 1000°C [Hojo et al., 1978] and the overall chemical reaction (Eq. (9-37)) is very fast, the nucleation rate constant was not affected by reaction temperature since the nucleation takes place rapidly once the gaseous reactants reached the temperature of 1000°C in this work. The experimental results [Ryu et al., 2008c] showing the complete consumption of WCl_6 even at a low temperature of 1200°C supports the simulation approach. In addition, the products experimentally obtained in the range of the reaction temperature from 1200°C to 1400°C were mostly WC. On the other hand, the growth rate constant decreased with a decrease in the reaction temperature from 1400°C to 1200°C. This indicates that the particle size was mainly affected by the dependence of the growth rate on the reaction temperature.

The temperature dependence of the growth rate constant is expressed by the following Arrhenius equation:

$$k_g = 3.53\times10^{-3}\exp\left(-\frac{10202}{T}\right) \tag{9-53}$$

which gives an apparent activation energy value of 84.8 (kJ/mol), as shown in Figure 9.32. This result must be considered approximate because experimental data were obtained only at three reaction temperatures.

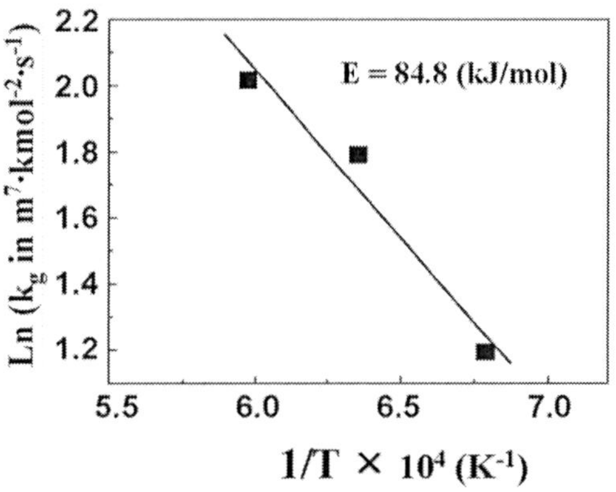

Figure 9.32. Arrhenius plot of the calculated apparent activation energy of the growth rate. [Adapted from Ryu et al. (2010c)].

9.6. Comparison between the Simulation Results of SiO_2 and WC Nanoparticles

The direct determination of the actual kinetics of nucleation and growth of nanoparticles synthesized in the gas phase is a very complicated task. It may include the measurement of the gas-phase concentration of products and precursors inside the reactor by means of in-situ spectroscopic techniques. Most of the previous modeling work on the gas-phase synthesis of nanoparticles involves one or several parameters fitted to yield agreements with the available experimental information such as average particle size, which is the most common and important parameter to be determined experimentally. In this laboratory, the simulation of the synthesis of SiO_2 nanoparticles in a flame reactor was also carried out. Simplified expression for the rate of nucleation and growth were proposed and two rate constants were adjusted as described above [Ji et al., 2007].

Although this fitting approach is empirical and the operating conditions (i.e. reaction temperature, residence time, etc.) of the synthesis processes of SiO_2 and WC nanoparticles are different, it is of interest to compare the predicted rates of nucleation and growth of particles in both processes. As a result, an order of magnitude comparison between both rate processes was performed. Since the average particle size was the target variable that was adjusted by fitting the rate constants, the comparison between the predicted average particle sizes in both synthesis processes will be meaningful. The average particle sizes were 15.6 and 21.7 nm for SiO_2 and WC nanoparticles, respectively. The magnitudes of the nucleation and growth rates of nanoparticles in both processes were in the order of 10^{18} $\#\cdot m^{-3}$ s^{-1} and 10^{-7} m s^{-1}, respectively. This result correlates well with the fact that both materials SiO_2 and WC have high melting points. This finding indicates that the proposed fitting approach is representing reasonable the actual phenomena taking place in both synthesis processes.

The chemical vapor synthesis (CVS) of WC nanopowder from tungsten hexachloride in a tubular reactor was simulated by a two-dimensional multiphase computational fluid dynamics (CFD) model. The nucleation process was described by the reaction of the precursor vapors in the vapor phase. The particle growth rate was expressed by incorporating the combined effects of surface reaction and mass transfer, but in this system the former was the rate-controlling step. A parametric analysis was used to extract the nucleation rate constant (k_n) and growth rate constant (k_g), because of the difficulty involved in their experimental measurements. Experimental results obtained for the synthesis of WC nanopowder from WCl_6 with H_2 and CH_4 in a tubular reactor system were used to validate the model. The combination of nucleation rate and growth rate constants that yielded the best agreement with experimental data was determined.

CONCLUSION

A mathematical modeling technique based on computational fluid dynamics (CFD) for the vapor-phase synthesis of nanopowders was developed. The technique incorporates an Eulerian-Eulerian approach (Mixture Model) to solve the three-dimensional turbulent governing equations of the vapor phase and the continuity equation for the particulate phase. The particle size distribution (PSD) of the nanopowders is described by the population balance model (PBM) which incorporates homogeneous nucleation, vapor condensation, surface reaction, and coagulation. The technique has been used to simulate the synthesis of aluminum, silica, and tungsten carbide nanopowders. In the case of aluminum nanopowder, the model is capable of predicting the experimental average particle size without the need of adjustable parameters. For silica nanopowder, a single value of the collision efficiency factor is sufficient to reproduce the magnitude as well as the variations of the average particle diameter with different experimental conditions.

To the best of the author's knowledge, this is the first application of a modeling technique that combines CFD, a multiphase model (Mixture Model) and a population balance model for the vapor-phase synthesis of aluminum and silica nanopowders. Overall, this technique has proved to be a promising tool for the design and scale up of industrial-scale vapor-phase synthesis reactors.

ACKNOWLEDGMENTS

During the years of my research and development of the chemical vapor synthesis process, I have been fortunate to work and collaborate with a number of wonderful people.

- Coworkers of my work discussed in each section were mainly my current and former students and postdoctoral research associates, who were more than just coworkers but really members of my extended family.
- Miguel Olivas-Martinez carefully read Chapter 9 and made many useful suggestions and he has done other outstanding work in my laboratory as a graduate student.
- Sohini PalDey started and established a firm foundation of my work on Chemical Vapor Synthesis research by her tireless work overcoming some serious obstacles. Taegong Ryu accomplished the largest amount of work on the subject by his hard work, initiative and original ideas. He was also a consummate team worker while in my laboratory, collaborating and forming teams with people around him when he came up with new ideas.
- Professor Zhigang Zak Fang of the University of Utah was instrumental in collaborating with me on the subjects of tungsten carbide nanopowder synthesis and hydrogen-storage materials development discussed in this book.
- I also had great external collaborators on the subject, including Dr. Hee Dong Jang and Dr. Kang-In Rhee of Korea Institute of Geoscience and Mineral Resources (KIGAM), Daejeon, Korea, Professor Insoo Kim of Dong-A University, Busan, Korea, and Professor Chang-Whan Won of Choongnam National University, Daejeon, Korea, among others.
- TMS has given me encouragements and motivation by the recognitions it has bestowed on my coworkers and me, which include the 1999 Extraction and Processing Science Award based on our work on CVS of intermetallic compounds discussed in Chapter 4.
- Last but not least, I wish to acknowledge the following funding agencies and organizations that have supported my research on the subject of nanopowder synthesis: Korea Institute of Geoscience and Mineral Resources (KIGAM); Chrysalis Technologies Inc.; U.S. Department of Interior; U.S. Department of Energy (several large grants); the State of Utah; the University of Utah Funding Incentive Seed Grant Committee; and the University of Utah Research Committee.

References

Adam J., R. Drumm, G. Klein and M. Veith 2008, Milling of Zirconia Nanoparticles in a Stirred Media Mill, *J. Am. Ceram. Soc.*, 91, 2836-43.

Andreasen A. 2008, Hydrogenation properties of Mg-Al alloys, *Int. J. Hydrog.* Energy, 33, 7489-97.

ANSYS, FLUENT 6.3 User's Guide: Lebanon, NH, 2006.

Arriagada F. J. and K. Osseo-Asare 1999, Synthesis of Nanosize Silica in a Nonionic Water-in-Oil Microemulsion: Effects of the Water/Surfactant Molar Ratio and Ammonia Concentration, *J. Colloid Interface Science,* 211, 210-220.

ASM International 1992, ASM handbook, Alloy phase diagrams. ASM International: *Materials Park*, OH, p 2-48, 2-282.

Axelbaum R.L. 1997, Developments in sodium/halide flame technology for the synthesis of unagglomerated non-oxide nanoparticles, in Proc. of the Joint NSF-NIST Conference on Nanoparticles: Synthesis, Processing into Functional Nanostructures and Characterization (May 12-13, Arlington, VA).

Bachtold A., P. Hadley, T. Nakanishi and C. Dekker 2001, Logic circuits with carbon nanotube transistors, *Science,* 294, 1317-20.

Balema V.P., K.W. Dennis and V.K. Pecharsky 2000, Rapid solid-state transformation of tetrahedral $[AlH_4]^-$ into octahedral $[AlH_6]^{3-}$ in lithium aluminohydride, *Chem. Commun.*, 17, 1665-66.

Ban, Z.-G. and L. L. Shaw 2002, Synthesis and processing of nanostructured WC-Co materials. *J. Mater. Sci.,* 37, 3397-03.

Barksdale J. 1966, Titanium Its Occurrence, Chemistry and Technology; Ronald Press Co.: New York, NY, pp 213-223, pp 400-79.

Becker M. F., J. R. Brock, H. Cai, D. E. Henneke, J. W. Keto, J. Lee, W. T. Nichols and H. D. Glicksman 1998, Metal nanoparticles generated by laser ablation, *Nanostructured Materials*, 10, 853-863.

Berggren A., A.E.C. Palmqvist and K. Holmberg 2005. Surfactant-templated mesostructured materials from inorganic silica. *Soft Matter* 1, 219-26.

Berndt C. C. and E. J. Lavernia 1998, Thermal spray processing of nanoscale materials—A conference report with extended abstract, *J. Thermal Spray Technol.*, 7, 411-440.

Bickmore C. R., K. F. Waldner, D. R. Treadwell and R. M. Laine 1996, Ultrafine spinel powders by flame spray pyrolysis of a magnesium aluminum double alkoxide. *J. Am. Ceram. Soc.,* 79, 1419-23.

Bickmore C. R., K. F. Waldner, R. Baranwal, T. Hinklin, D. R. Treadwell and R. M. Laine 1998, Ultrafine Titania by Flame Spray Pyrolysis of a Titanatrane Complex. *J. Eur. Ceram. Soc.*, 18, 287-97.

Bird R. B., W. E. Stewart and E. N. Lightfoot 2002, Transport Phenomena, 2nd edition, Wiley, New York.

Bless S. J., K. Tarcza, R. Chau, E. Taleff and C. Persad 2006, Dynamic fracture of tungsten heavy alloys, *Int. J. Impact Eng.*, 33.,100-08.

Bogdanovic B. and M. Schwickardi 1997, Ti-doped alkali metal aluminium hydrides as potential novel reversible hydrogen storage materials, *J. Alloys Compd.*, 253-254, 1-9.

Bogdanovic B., M. Felderhoff, M. Germann, M. Härtel, A. Pommerin, F. Schüth, C. Weidenthaler and B. Zibrowius 2003, Investigation of hydrogen discharging and recharging processes of Ti-doped $NaAlH_4$ by X-ray diffraction analysis (XRD) and solid-state NMR spectroscopy, *J. Alloys Compd.* 350, 246-255.

Bolsaitis P. P., J. F. McCarthy, G. Mohiuddin and J. F. Elliot 1987, Fomation of Metal Oxide Aerosols for Conditions of High Supersaturation, *Aerosol. Sci. Tech.,* 6, 225-46.

Bondioli F., A. M. Ferrari, C. Leonelli, C. Siligardi and G. C. Pellacani 2001, Microwave-Hydrothermal Synthesis of Nanocrystalline Zirconia Powders, *J. Am. Ceram. Soc.*, 84, 2728-30.

Bouaricha S., J.P. Dodelet, D. Guay, J. Huot, S. Boily and R. Schulz 2000, Hydriding behavior of Mg-Al and leached Mg-Al compounds prepared by high-energy ball-milling, *J. Alloys Compd.* 297, 282-293.

Brousseau P. and C. J. Anderson 2002, Nanometric aluminum in explosives, Propellants, *Explos., Pyrotech.*, 27, 300-06.

Brousseau P. and M .D. Cliff 2001, The Effect of Ultrafine Aluminum Powder on the Detonation Properties of Various Explosives, 32nd Int. Annual Conference of ICT, Karlsruhe, Germany, July 3-6, pp 37/1-14.

Calcote, H.F. and D.G. Keil 1997, Combustion synthesis of silicon carbide powder, in Proc. of the Joint NSF-NIST Conference on Nanoparticles: Synthesis, Processing into Functional Nanostructures and Characterization (May 12-13, Arlington, VA).

Cao G. 2004, Nanostructures and Nanomaterials: Synthesis, Properties and Applications, Imperial College Press, London.

Carroll D. F. 1999, Sintering and microstructural development in WC/Co-based alloys made with superfine WC powder, *Int. J. Refract. Met. Hard Mater.* 17, 123-33.

Caruso F., R. A. Caruso and H. Mohwald 1998, Nanoengineering of inorganic and hybrid hollow spheres by colloidal templating, *Science,* 282, 1111-14.

Chang W., G. Skandan, H. Hahn, S. C. Danforth and B. H. Kear 1994, Chemical vapor condensation of nanostructured ceramic powders, *Nanostructured Materials*, 4, 345-51.

Chang H. W., I. W. Lenggoro, K. Okuyama and H. D. Jang 2006, Flame spray pyrolysis for preparing red-light-emitting, submicron-sized luminescent strontium titanate particles. *Japanese Journal of Applied Physics Part 1-Regular Papers Brief Communications and Review Papers*, 45, 967-73.

Chazelas C., J. F. Coudert, J. Jarrige and P. Fauchais 2007, Synthesis of nanometer alumina particles by plasma transferred arc alternative, *J. Eur. Ceram. Soc.,* 27, 947-50.

Chervin C. N., B. J. Clapsaddle, H. W. Chiu, A. E. Gash, J. H. Satcher Jr., and S. M. Kauzlarich 2006, Role of Cyclic Ether and Solvent in a Non-Alkoxide Sol–Gel Synthesis of Yttria-Stabilized Zirconia Nanoparticles, *Chem. Mater.*, 18, 4865-74.

Cho K., H. Chang, D. S. Kil, J. Park, H. D. Jang and H. Y. Sohn 2009, Mechanisms of the Formation of Silica Particles from Precursors with Different Volatilities by Flame Spray Pyrolysis, *Aerosol Sci. Technol.*, 43, 911-20. http://www.informaworld.com/smpp/content~db=all~content =a912455996~frm=titlelink?words=sohnandhash=2532445632.

Choi J.W. 2009, Chemical vapor synthesis and characterization of metallic nanopowders relevant to hydrogen storage materials, Ph.D. dissertation, University of Utah, Salt Lake City, Utah.

Choi S. U. S., Z. G. Zhang, W. Yu, F. E. Lockwood and E. A. Grulke 2001, Anomalous thermal conductivity enhancement in nanotube suspensions, *Applied Physics Letters*, 79, 2252-54.

Choi S. I., J. S. Nam, C. M. Lee, S. S. Choi, J. I. Kim, J. M. Park and S. H. Hong 2006, High purity synthesis of carbon nanotubes by methane decomposition using an arc-jet plasma, *Curr. Appl. Phys.*, 6, 224-29. doi:10.1016/j.cap.2005.07.045.

Choi Y. J., J. Lu, H. Y. Sohn and Z. Z. Fang 2008, Hydrogen Storage Properties of the Mg-Ti-H System Prepared by High-Energy-High-Pressure Reactive Milling, *J. Power Sources*, 180, 491–97. http://dx.doi.org/10.1016/j.jpowsour.2008.02.038.

Choi Y. J., J. W. Choi, H. Y. Sohn, T. Ryu, K. S. Hwang and Z. Z. Fang 2009a, Chemical Vapor Synthesis of Mg/Ti Nanopowder Mixture as a Hydrogen Storage Material, *Int. J. Hydrogen Energy*, 34, 7700-06; http://dx.doi.org/10.1016/j.ijhydene.2009.07.033.

Choi Y. J., J. Lu, H. Y. Sohn, Z. Z. Fang and E. Ronnebro 2009b, Effect of Milling Parameters on the Dehydrogenation Properties of the Mg-Ti-H System, *J. Phys. Chem.* C, 113 (44), 19344–50. http://pubs.acs.org/ articlesonrequest/AOR-YJS2Efu5FDx5QmIjGURC.

Choi J. W., H. Y. Sohn, Y. J. Choi, and Z. Z. Fang 2010, Chemical Vapor Synthesis and Characterization of Aluminum Nanopowder, *J. Power Sources*, 195, 1463–71.

Chraska T., A. H. King and C. C. Berndt 2000, On the size-dependent phase transformation in nanoparticulate zirconia, *Mater. Sci. Eng.,* 286, 169-78.

Chu J. C. S., J. Breslin, N. S. Wang and M. C. Lin 1991, Relative stabilities of tetramethylorthosilicate and tetraethylorthosilicate in the gas phase, *Mater. Lett.*, 12, 179-84.

Collins P. G., M. S. Arnold and Ph. Avouris 2001, Engineering Carbon Nanotubes and Nanotube Circuits Using Electrical Breakdown, *Science,* 292, 706-09.

Crivello J.-C., T. Nobuki, S. Kato, M. Abe and T. Kuji 2007, Hydrogen absorption properties of the γ-$Mg_{17}Al_{12}$ phase and its Alricher domain, *J. Alloys Compd.*, 446-447, 157-61.

Crivello J.-C., T. Nobuki and T. Kuji 2009, Improvement of Mg-Al alloys for hydrogen storage applications, *Int. J. Hydrog. Energy,* 34, 1937-43.

Cullity B. D. 1978, Elements of X-ray diffraction, 2nd edition, Addison-Wesley, Reading, MA, 99-106, 415-17.

Cutler R. A., J. R. Reynolds and A. Jones 1992, Sintering and Characterization of Polycrystalline Monoclinic, Tetragonal, and Cubic Zirconia, *J. Am. Ceram. Soc.*, 75, 2173-83.

Dabadie T. , A. Ayral, C. Guizard, L. Cot, C. Lurin, W. Nie and D. Rioult 1995, Synthesis of inorganic gels in a lyotropic liquid crystal medium. I. Synthesis of silica gels in lamellar phases obtained from non-ionic surfactants, *J. Sol-Gel. Sci. Technol.* 4, 107-16.

Dastafani J.D. 1989, Advances in intermetallics, *Adv. Mater. Processes*, 135 (2), 37-41.

Davidson C. F., G. B. Alexander and M. E. Wadsworth 1979, Catalytic effect of cobalt on the carburization kinetics of tungsten, *Metall. Trans. A, 10A,* 1059 –69.

Derycke V., R. Martel, J. Appenzeller and Ph. Avouris 2001, Carbon Nanotube Inter- and Intramolecular Logic Gates, *Nano Letters*, 1, 453–56.

Eastman J. A., S. U. S. Choi, S. Li, W. Yu and L. J. Thompson 2001, Anomalously increased effective thermal conductivities of ethylene glycol-based nanofluids containing copper nanoparticles, *Applied Physics Letters*, 78, 718-20.

Edelstein A.S. and R.C. Cammarata, eds. 1998, Nanomaterials: Synthesis, Properties, and Applications, Institute of Physics Publishing.

Ehrman S. H. 1999, Effect of Particle Size on Rate of Coalescence of Silica Nanoparticles, J. Colloid Interf. Sci., 213, 258-61.

Ehrman S. H., S. K. Friedlander and M. R. Zachariah 1998, Characteristics of SiO_2/TiO_2 nanocomposite particles formed in a premixed flat flame, *J. Aerosol. Sci.*, 29, 687-706.

Fait M. J. G., H.-J. Lunk, M. Feist, M. Schneider, J. N. Dann and T. A. Frisk 2008, Thermal decomposition of ammonium paratungstate tetrahydrate under non-reducing conditions: Characterization by thermal analysis, X-ray diffraction and spectroscopic methods, *Thermochim. Acta*, 469, 12-22.

Fang Z. and J. W. Eason 1995, Study of Nanostructured WC-Co Composites, *Int. J. Refract. Met. Hard Mater.*, 13, 297-303.

Fang Z., P. Maheshwari, X. Wang, H. Y. Sohn, A. Griffo and R. Riley 2005, An Experimental Study of Sintering of Nanocrystalline WC-Co Powders, *Int. J. Refractory Metals and Hard Materials, 23,* 249-57. doi:10.1016/j.ijrmhm.2005.04.014.

Fang Z. Z., X. Wang, T. Ryu and H.Y. Sohn 2009, Synthesis, Sintering, and Mechanical Properties of Nanocrystalline Cemented Tungsten Carbide – A Review, *Int. J. Refractory Metals and Hard Materials*, 27, 288–99. doi:10.1016/j.ijrmhm.2008.07.011.

Fichtner M. and O. Fuhr 2002, Synthesis and structures of magnesium alanate and two solvent adducts, *J. Alloys Compd.*, 345, 286-96.

Fichtner M., O. Fuhr and O. Kircher 2003, Magnesium alanate - a material for reversible hydrogen storage?, *J. Alloys Compd.*, 356-357, 418-22.

Fitzsimmons M. and V. K. Sarin 1995, Comparison of WCl_6-CH_4-H_2 and WF_6-CH_4-H_2 systems for growth of WC coatings, *Surface and Coatings Technology*, 76-77, Part 1, 250-55. doi:10.1016/0257-8972(95)02568-5.

Fluent 6.3.26, 2003, User's Guide. Fluent Inc., Lebanon, NH.

Fluent 6 Population Balance Model Manual 2004, Fluent Inc., Lebanon, NH.

Fossdal A., H. W. Brinks, M. Fichtner and B. C. Hauback 2005, Thermal decomposition of $Mg(AlH_4)_2$ studied by in situ synchrotron X-ray diffraction, *J. Alloys Compd.*, 404-406, 752-56.

French G. J. and F. R. Sale 1985, The hydrogen reduction of cobalt-tungsten mixed oxides, *J. Mater. Sci.*, 20, 1291-300.

Friedlander S. K. 2000, Smoke, Dust, and Haze, Oxford University Press, New York.

Fu L., L. H. Cao and Y. S. Fan 2001, Two-step synthesis of nanostructured tungsten carbide-cobalt powders, *Scripta Mater.*, 44, 1061-68. doi:10.1016/S1359-6462(01)00668-6.

Fukumasa O. and T. Fujiwara 2003, Rapid synthesis of ferrite particles from powder mixtures using thermal plasma processing, *Thin Solid Films*, 435, 33-38.

Gao L. and B.H. Kear 1995, Low temperature carburization of high surface area tungsten powders, *Nanostruct. Mater.* 5, 555-69.

Gao Y., J. Zhao, Y. Zhu, S. Ma, X. Su and Z. Wang 2006, Wet chemical process of rod-like tungsten nanopowders with iron (II) as reductive agent, *Mater. Lett.*, 60, 3903-05.

Gell M. 1995, Application Opportunities for Nanostuctured Materials and Coatings, *Mater. Sci. Eng. A,* 204, 246-51.

German R. M., J. Ma, X. Wang and E. Olevsky 2006, Processing model for tungsten powders and extension to nanoscale size range, *Powder. Metall.*, 49, 19-27.

Gibson I. R. and J. T. S. Irvine 2001, Qualitative X-Ray Diffraction Analysis of Metastable Tetragonal (t') Zirconia, *J. Am. Ceram. Soc.,* 84, 615-18.

Giesche H. 1994, Synthesis of monodispersed silica powders II. Controlled growth reaction and continuous production process, *J. Eur. Ceram. Soc.*, 14, 205-14.

Gimblett F. G. R., A. Hussain, and K. S. W. Sing 1988, Thermal and Related Studies of Some Basic Zirconium Salts, *J. Therm. Anal.*, 34, 1001-13.

Gleiter H. 1989, Nanocrystalline materials, *Progress in Materials Science*, 33, 223-315.

Gleiter H. 1992, Materials with Ultrafine Microstructures: Retrospectives and Perspectives *Nanostruct. Mater.* 1, 1-19.

Gleiter H. 2000, Nanostructured materials: basic concepts and microstructure, *Acta mater.*, 48, 1-29.

Gostishchev V. V., V. F. Boiko, L. A. Klimova and L P. Metlitskaya 2006, Physicochemical aspects of tungsten powder production by reduction of tungsten compounds by aluminum in ionic melts, *Theor. Found. Chem. Eng.*, 40, 546-49.

Graetz J. and J. J. Reilly 2005, Decomposition Kinetics of the AlH_3 Polymorphs, *J. Phys. Chem.* B, 109, 22181-85.

Graham D. D., L. F. Culnane, M. Sulic, C. M. Jensen and I.M. Robertson 2007, Ti EELS standards for identification of catalytic species in $NaAlH_4$ hydrogen storage materials, *J. Alloys Compd.* 446-447, 255-259.

Greenwood N. N. and A. Earnshaw 1998, Chemistry of the elements, second ed., Butterworth-Heinemann: Oxford [England], Boston, pp. 32-64.

Gremaud R., A. Borgschulte, W. Lohstroh, H. Schreuders, A. Züttel, B. Dam and R. Griessen 2005, Ti-catalyzed $Mg(AlH_4)_2$: A reversible hydrogen storage material, *J. Alloys Compd.*, 404-406, 775-78.

Grochala W, and P. Edwards 2004, Thermal Decomposition of the Non-Interstitial Hydrides for the Storage and Production of Hydrogen, *Chem. Rev.*,104, 1283-316.

Herzler J., J. A. Manion and W. Tsang 1997, Single-Pulse Shock Tube Study of the Decomposition of Tetraethoxysilane and Related Compounds, *J. Phys. Chem. A,* 101, 5500-08.

Hinds W. C. 1999, Aerosol Technology. Wiley, New York.

Hojo J., T. Oku and A. Kato 1978, Tungsten carbide powder produced by the vapor phase reaction of the WCl_6-CH_4-H_2 system, *J. Less Common Metals,* 59, 85-95. doi:10.1016/0022-5088(78)90114-5

Hopwood J. D. and S. Mann 1997, Synthesis of Barium Sulfate Nanoparticles and Nanofilaments in Reverse Micelles and Microemulsions, *Chem. Mater.*, 9, 1819–28.

Howard C. J. and R. J. Hill 1991, The Polymorphs of Zirconia: Phase Abundance and Crystal Structure by Rietveld Analysis of Neutron and X-Ray Diffraction Data, *J. Mater. Sci.*, 26, 127-34.

Hu M. Z. C., R. D. Hunt, E. A. Payzant and C. R. Hubbard 1999, Nanocrystallization and Phase Transformation in Monodispersed Ultrafine Zirconia Particles from Various Homogeneous Precipitation Methods, *J. Am. Ceram. Soc.*, 82, 2313-20.

Iler R. K. 1979, The Chemistry of Silica, Wiley, New York.

Inagaki S., S. Guan, T. Ohsuna and O. Terasaki 2002, An ordered mesoporous organosilica hybrid material with a crystal-like wall structure, *Nature*, 416, 304-07.

Iniguez J. and T. Yildirim 2005, First-principles study of Ti-doped sodium alanate surfaces, *Appl. Phys. Lett.*, 86, 103109_1-193109_3.

Ishikawa H., Otsuka K.-I., M. Nitta and Y. Yonehana 1990, Preparation of ultrafine metallic particles by hydrogen reduction of chloride vapors, *Kawasaki Steel Tech. Rept.*, No. 23, 41-47.

Ivanov G. V. and F. Tepper 1997, Activated Aluminum as a Stored Energy Source for Propellants, inChallenges in Propellants and Combustion 100 Years After Nobel, K. K. Kuo (Eds), Begell House, New York, p. 636.

Jang H. D. 1999, Generation of Silica Nanoparticles from Tetraethylorthosilicate (TEOS) Vapor in a Diffusion Flame, *Aerosol Science and Technology*, 30,477-88.

Jang H.D. 2001, Experimental study of synthesis of silica nanoparticles by a bench-scale diffusion flame reactor, *Powder Technology*, 119, 102-18.

Jang H. D., C. M. Seong, Y. J. Suh, H. C. Kim and C. K. Lee 2004, Synthesis of lithium-cobalt oxide nanoparticles by flame spray pyrolysis. *Aerosol Sci. Technol.*, 38, 1027-32.

Jang H. D., H. Chang, Y. Suh and K. Okuyama 2006, Synthesis of SiO_2 nanoparticles from sprayed droplets of tetraethylorthosilicate by the flame spray pyrolysis. *Current Applied Physics* 6, pp. e110-e113.

Jang H. D., D. S. Kil, H. K. Chang, K. Cho and H. Y. Sohn 2007, Silica Nanopowder from Waste Silicon Sludge, Proceedings of the International Symp. of Resource Recycling, Jeju Island, Korea, Nov. 28-29.

Jensen C.M., R. Zidan, N. Mariels, A. Hee and C. Hagen 1999, Advanced titanium doping of sodium aluminum hydride, *Int. J. Hydrog. Energy* 24, 461-65.

Ji Y., H. Y. Sohn, H. D. Jang, B. Wan and T. A. Ring 2007, Computational Fluid Dynamic Modeling of Flame Reaction Process for Silica Nanopowder Synthesis from Tetraethylorthosilicate (TEOS), *J. Am. Ceram. Soc.*, 90, 3838-45. http://www3.interscience.wiley.com/cgi-bin/fulltext/117955367/PDFSTART

Jossen R., S. E. Pratsinis, W. J. Stark and L. Madler 2005, Criteria for flame-spray synthesis of hollow, shell-like, or inhomogeneous oxides, *J. Am. Ceram. Soc.,* 88, 1388-93.

Kang X.D., P. Wang, X.P. Song, X.D. Yao, G.Q. Lu and H.M. Cheng 2006, Catalytic effect of Al_3Ti on the reversible dehydrogenation of $NaAlH_4$, *J. Alloys Compd.* 424, 365-69.

Kawakami N., Y. Fukumoto, T. Kinoshita, K. Suzuki and K. Inoue 2000, Preparation of highly porous aerogel thin film by supercritical drying, Jpn, *J. Appl. Phys.*, Part 2-Letters, 39, L182-L184.

Kear B. H. and L. E. McCandlish 1993, Chemical processing and properties of nanostructured WC-Co materials, *Nanostruct. Mater.*, 3, 19-30.

Kearns M. 2004, Development and applications of ultrafine aluminium powders, *Mater. Sci. Eng.* A 375, 120–26.

Kecskes L. J., K. C. Cho, R. J. Dowding, B. E. Schuster, R. Z. Valiev, and Q. Wei 2007, Grain size engineering of bcc refractory metals: Top-down and bottom-up - Application to tungsten, *Mater. Sci. Eng. A*, 467, 33-43.

Kelly C. M., D. Garg and P. N. Dyer 1992, Kinetics of chemical vapor deposition of tungsten carbide, *Thin Solid Films*, 219, 103-08. doi:10.1016/0040-6090(92)90729-U

Kelsall R. 2005, Nanoscale Science and Technology, Wiley.

Keren K., R. S. Berman, E. Buchstab, U. Sivan and E. Braun 2003, DNA-templated carbon nanotube field-effect transistor, *Science,* 302, 1380–82.

Kilian A.and T. F. Morse 2001, A Novel Aerosol Combustion Process for the High Rate Formation of Nanoscale Oxide Particles, *Aerosol Sci. Technol.*, 34, 227-35.

Kim Y.W. 1994, Ordered intermetallic alloys, Part III: Gamma titanium aluminides, *JOM* 46 (7), 30–39.

Kim J. C. and B. K. Kim 2004, Synthesis of nanosized tungsten carbide powder by the chemical vapor condensation process. *Scripta Mater.*, 50, 969-72. doi:10.1016/ j.scriptamat.2004.01.015.

Kim Y., E.-K. Lee, J.-H. Shim, Y. W. Cho and K. B. Yoon 2006, Mechanochemical synthesis and thermal decomposition of $Mg(AlH_4)_2$, *J. Alloys Compd.*, 422, 283-87.

Kiminami R. H. G. 1990, The Monoclinic-Tetragonal Phase Transformation of Zirconia in the System ZrO_2–Fe_3O_4, *J. Mater. Sci. Lett.*, 9, 373-74.

Kishida M., K. Umakoshi, J.-I. Ishiyama, H. Nagata and K. Wakabayashi 1996, Hydrogenation of carbon dioxide over metal catalysts prepared using microemulsion, *Catalysis Today*, 29, 355-59.

Kodas T. T. and M. J. Hampden-Smith 1999, Aerosol processing of materials, Wiley-VCH: New York, pp. 97-100; 124-27.

Komiya K., N. Morisaku, Y. Shinzato, K. Ikeda, S. Orimo, Y. Ohki, K. Tatsumi, H. Yukawa and M. Morinaga 2007, Synthesis and dehydrogenation of $M(AlH_4)_2$ (M = Mg, Ca), *J. Alloys Compd.*, 446-47, 237-41.

Kong, L.B., J. Ma, W. Zhu and O.K. Tan 2001, Preparation of PMN powders and ceramics via a high-energy ball milling process, *J. Mater. Sci. Lett.,* 20, 1241-43.

Kortan A. R., R. Hull, R. L. Opila, M. G. Bawendi, M. L. Steigerwald, P. J. Carroll, L. E. Brus 1990, Nucleation and growth of cadmium selendie on zinc sulfide quantum crystallite seeds, and vice versa, in inverse micelle media, *J. Am. Chem. Soc.*, 112, 1327–32.

Krakhmalev P. V., J. Sukumaran and A. Gåård 2007, Effect of microstructure on edge wear mechanisms in WC-Co, *Int. J. Refract. Met. Hard Mater.,* 25, 171-78.

Kroll W. J. 1940, The production of ductile titanium, *Trans. Electrochem. Soc.*, 78, 35-47.

Kruis J.-O. and H. Fissan 1999, Nano-Process Technology for Synthesis and Handling of Nanoparticles, KONA, 17, 130-39.

Kubaschewski O. and C.B. Alcock 1979, Metallurgical Thermochemistry, 5th ed., Pergamon Press, Elmsford, NY, pp. 358-77.

Kung H. H. and E. I. Ko 1996, Preparation of oxide catalysts and catalyst supports — a review of recent advances, *Chem. Eng. J. and Biochem. Eng. J.*, 64, 203-14.

Kyoi D., T. Sato, E. Ronnebro, N. Kitamura, A. Ueda, M. Ito, S. Katsuyama, S. Hara, D. Noréus, and T. Sakai 2004, A new ternary magnesium-titanium hydride Mg7TiHx with hydrogen desorption properties better than both binary magnesium and titanium hydrides, *J. Alloys Compd.*, **372**, 213-17.

Kyprianidou-Leodidou T., W. Caseri, U. W. Suter 1994, Size Variation of PbS Particles in High-Refractive-Index Nanocomposites, *J. Phys. Chem.*, 98, 8992–97.

Lalena J. N. 2008, Inorganic Materials Synthesis and Fabrication, Wiley.

Lamprey H. and R. L. Ripley 1962, Ultrafine tungsten and molybdenum powders, *J. Electrochem. Soc.*, 109, 713-15.

Lassner E. and W.-D. Schubert 1999, Tungsten-Properties, Chemistry, Technology of the Element, Alloys, and Chemical Compounds, New York, Kluwer Academic/Plenum Publishers, p. 215-25.

Leclercq G., M. Kamal, J. M. Giraudon, P. Devassine, L. Feigenbaum, L. Leclercq, A. Frennet, J. M. Bastin, A. Löfberg, S. Decker and M. Dufour 1996, Study of the Preparation of Bulk Powder Tungsten Carbides by Temperature Programmed Reaction with CH_4+H_2 Mixtures, *J. Catal.,* 158, 142-69. doi:10.1006/jcat.1996.0015

Lee G.-H. and S. Kang 2006, Sintering of nano-sized WC–Co powders produced by a gas reduction–carburization process, *J. Alloy Comp.*, 419, 281-89. doi:10.1016/ j.jallcom.2005.09.060.

Lee B. W., S. Oh and M. Choi 2001, Simulation of growth of nonspherical silica nanoparticles in a premixed flat flame, *Aerosol Sci. Tech.*, 35, 978-89.

Lee M. H., K. Cho, A. P. Shah and P. Biswas 2005, Nanostructured sorbents for capture of cadmium species in combustion environments, *Environ. Sci. Technol.*, 39, 8481-89.

Lee S. H., S.-M. Oh and D.-W. Park 2007, Preparation of silver nanopowder by thermal plasma, *Mater. Sci. Eng*., C, 27, 1286-90.

Lefebvre, A. H. 1989, Atomization and Sprays, Taylor and Francis, London.

Lehtinen K. E. J., R. S. Windeler and S. K. Friedlander 1996, A Note on the Growth of Primary Particles in Agglomerate Structures by Coalescence, *J. Colloid Interf. Sci.*, 182, 606-08.

Lei H., Y.-J. Tang, J.-J. Wei, J. Li, X.-B. Li, and H.-L. Shi 2007, Synthesis of tungsten nanoparticles by sonoelectrochemistry, Ultrason. *Sonochem.*, 14, 81-83.

Lessard P., F. Beaupre and P. Brousseau 2001, Burn Rate Studies of Composite Propellants Containing Ultra-Fine Metals, 32nd Int. Annual Conference of ICT, Karlsruhe, Germany, 3 - 6 July 2001, p. 88/1.

Li C. and M. Li 2002, UV Raman Spectroscopic Study on the Phase Transformation of ZrO_2, Y_2O_3–ZrO_2 and SO_4^{2-}/ZrO_2, *J. Raman Spectrosc.*, 33, 301-08.

Li S., V. T. John, C. O'Connor, V. Harris, and E. Carpenter 2000, Cobalt-ferrite nanoparticles: Structure, cation distributions, and magnetic properties, *J. Appl. Phys*. 87, 6223-25.

Liu C. T. and K. S. Kumar 1993, Ordered intermetallic alloys, Part I: Nickel and iron aluminides, *JOM* 45 (5), 38–44.

Liu C.T., C.L. White, C.C. Koch and E.H. Lee 1983, Preparation of ductile nickel aluminides for high temperature use, in High Temperature Materials Chemistry—II, vol. 83 (7), L.A. Munir, D. Cubicciotti (Eds.), Electrochemical Society, Pennington, NJ, pp. 32–41.

Liu, Y., J. Zhang, L. Yu, G. Jia, C. Jing and S. Cao 2004, Magnetic properties of nanocrystalline Fe-Co alloys by high energy milling, *J. Alloys Compd.*, 377, 202-06.

Lowe T. 2002, The revolution in nanometals, *Adv. Mater. Process*., 160, 63–65.

Lu Y., R. Ganguli, C. A. Drewien, M. T. Anderson, C. J. Brinker, W. Gong, Y. Guo, H. Soyez, B. Dunn, M. H. Huang and J. I. Zink 1997, Continuous formation of supported cubic and hexagonal mesoporous films by sol-gel dip-coating, *Nature*, 389, 364-68.

Lu J., Z. Z. Fang and H. Y. Sohn 2006, A New Li-Al-N-H System for Reversible Hydrogen Storage, *J. Phys. Chem. B*, 110, 14236-39.

Lu J., Y. J. Choi, Z. Z. Fang, H. Y. Sohn and E. Ronnebro 2009, Hydrogen Storage Properties of Nanosized MgH_2-0.1TiH_2 Prepared by Ultrahigh-Energy-High-Pressure Milling, *J. Am. Chem. Soc.*, 131 (43), 15843–52.

Madarász J., I. M. Szilágyi, F. Hange and G. Pokol 2004, Comparative evolved gas analyses (TG-FTIR, TG/DTA-MS) and solid state (FTIR, XRD) studies on thermal decomposition of ammonium paratungstate tetrahydrate (APT) in air, *J. Anal. Appl. Pyrolysis*, 7, 197-201.

Madler L. and S. E. Pratsinis 2002, Bismuth oxide nanoparticles by flame spray pyrolysis. *J. Am. Ceram. Soc.*, 85, 1713-18.

Madler L., H. K. Kammler, R. Mueller and S. E. Pratsinis 2002, Controlled synthesis of nanostructured particles by flame spray pyrolysis. *J. Aerosol Sci.*, 33, 369-89.

Magnussen B. F. and B. H. Hjertager 1976, On mathematical models of turbulent combustion with special emphasis on soot formation and combustion, In Sixteenth Symposium (International) on Combustion; The Combustion Institute, Pittsburgh, PA, 719-29.

Magnusson M. H., K. Deppert and J.-O. Malm 2000, Single-crystalline tungsten nanoparticles produced by thermal decomposition of tungsten hexacarbonyl, *J. Mater. Res.*, 15, 1564-69.

Malewar R, K. S. Kumar, B. S. Murty, B. Sarma and S. K. Pabi 2007, On sinterability of nanostructured W produced by high-energy ball milling, *J. Mater. Res.*, 22, 1200-06.

Manikama V. R., K. Y. Cheong and K. A. Razaka 2011, Chemical reduction methods for synthesizing Ag and Al nanoparticles and their respective nanoalloys, Mater. Sci. Eng. B, 176, 187-203.

Martel R., T. Schmidt, H. R. Shea, T. Hertel, and Ph. Avouris 1998, Single- and multi-wall carbon nanotube field-effect transistors, *Applied Physics Letters*, 73, 2447-49.

Mayo M. J., J. R. Seidensticker, D. C. Hague and A. H. Carim 1999, Surface chemistry effects on the processing and superplastic properties of nanocrystalline oxide ceramics, *Nanostruct. Mater.*, 11, 271-82.

McCandlish L. E. and R.S. Polizzotti 1989, Control of composition and microstructure in the Co-W-C system using chemical synthetic techniques, *Solid State Ionics*, 32-33, 795-801.

McCandlish, L. E., B. H. Kear and B. K. Kim 1992, Processing and properties of nanostructured WC-Co, *Nanostruct. Mater.*, 1, 119-24.

McGraw R. 1997, Description of Aerosol Dynamics by the Quadrature Method of Moments, *Aerosol Sci. Tech.*, 27, 255-65.

Medeiros F. F. P., S. A. De Oliveira, C. P. De Souza, A. G. P. Da Silva, U. U. Gomes and J. F. De Souza 2001, Synthesis of tungsten carbide through gas-solid reaction at low temperatures, *Mater. Sci. Eng., A,* 315, 58-62. doi:10.1016/S0921-5093(01)01214-X.

Mehrotra R.C. 1988, Synthesis and reactions of metal alkoxides, *J. Non. Cryst. Solids*, 100, 1-15.

Mench M. M., C. L. Yeh and K. K. Kuo 1998, Propellant Burning Rate Enhancement and Thermal Behavior of Ultra-Fine Aluminum Powders (Alex), 29th Int. Annual Conference of ICT, Karlsruhe, Germany, 30 June - 3 July 1998, p. 30/1.

Messing G. L., S.-C. Zhang and G. V. Jayanthi 1993, Ceramic Powder Synthesis by Spray Pyrolysis, *J. Amer. Ceram. Soc.*, 76, 2707–26.

Miclea C., C. Tanasoiu, A. Georghiu, C. F. Miclea and V. Tanasoiu 2004, Synthesis and piezoelectric properties of nanocrystalline PZT-based ceramics prepared by high energy ball milling process. *J. Mater. Sci.*, 39, 5431-34.

Minkler W. W. and E. F. Baroch 1981, Metallurgical Treatises, eds. J.K. Tien and J.F. Elliott, TMS, Warrendale, PA, p. 171.

Moriyoshi Y., M. Futaki, S. Komatsu and T. Ishigaki 1997, The preparation and characterization of ultrafine tungsten powder, *J. Mater. Sci. Lett.*, 16, 347-49.

Mueller R., R. Jossen, S. E. Pratsinis, M. Watson and M. K. Akhtar 2004, Zirconia Nanoparticles Made in Spray Flames at High Production Rates. *J. Am. Ceram. Soc.*, 87, 197-202.

Nersisyan H. H., H. I. Won, C. W. Won and J. H. Lee 2005, Study of the combustion synthesis process of nanostructured WC and WC–Co, *Mater. Chem. Phys.*, 94, 153-58. doi:10.1016/j.matchemphys.2005.04.024.

Nikklason 1987, G. A., Optical properties of gas-evaporated metal particles: effects of a fractal structure, *J. Appl. Phys.,* 62, 258-65.

Nishizawa T. and K. Ishida 1993, Co-Fe (Cobalt – Iron), in Phase Diagrams of Binary Iron Alloys, ed. by Hiroaki Okamoto, ASM International, Materials Park, OH 44073, pp. 93-101.

Oelerich W., T. Klassen and R. J Bormann 2001, Metal Oxides as Catalysts for Hydrogen Sorption of nanocrystalline Mg-based Alloys, *Alloys Compd.*, 315, 237-42.

Okamoto, Y., T. Koyano and A. Takasaki 1987, A magnetic study of sintering of ultrafine particles, *Japan J. Appl. Phys.*, 26, 1943-45.

Olivas-Martinez M., H. Y. Sohn and H. D. Jang 2008, Silica Nanopowder from Waste Silicon Sludge and CFD Modeling of the Flame Spray Pyrolysis Process, In REWAS 2008: Global Symposium on Recycling, Waste Treatment and Clean Technology; B. Mishra, C. Ludwig and S. Das; The Minerals, Metals and Materials Society: Warrendale, PA, pp 631-638.

Otsuka K.-I., H. Yamamoto and A. Yoshizawa 1984, Preparation of ultrafine particles of nickel, cobalt and iron by hydrogen reduction of chloride vapours, Nippon Kagaku Kaishi (*J. Chem. Soc. Japan, Chem. and Industrial Chem.*), No. 6, 869-78.

Outotec 2002, HSC Chemistry, Version 5.1, Finland.

PalDey S. 1995, Synthesis of Intermetallic Compounds by Vapor-Phase Coreduction of Metal Chlorides, Ph.D. dissertation, University of Utah.

Panda S. and S. E. Pratsinis 1995, Modeling the synthesis of aluminum particles by evaporation-condensation in an aerosol flow reactor, *Nanostruct. Mater.*, 5, 755-67.

Parr, H. and J. Feder 1973, Superconductivity in β-phase gallium, *Phys. Rev.,* 7, 166-81.

Paschen P., A. Anzinger and H. Feichtner 1990, Titanium – a techno-commercial evaluation of the reduction process, in Refractory Metals, eds. K. C. Lidell, D. R. Sadoway, and R. G. Bautista, TMS, Warrendale, PA, 7-13.

Patel H. E., S. K. Das, T. Sundararajan, A. S. Nair, B. George, and T. Pradeep 2003, Thermal conductivities of naked and monolayer protected metal nanoparticle based nanofluids: Manifestation of anomalous enhancement and chemical effects, *Applied Physics Letters*, 83, 2931-33.

Perez-Fontes S. E. and H. Y. Sohn 2011, Three Dimensional CFD-Population Balance Simulation of a Chemical Vapor Synthesis Reactor for Aluminum Nanopowder: Nucleation, Surface Growth and Coagulation, University of Utah, work to be published.

Petersson A. and J. Ågren 2004, Constitutive behaviour of WC–Co materials with different grain size sintered under load, *Acta Mater* 52:1847-58. doi:10.1016/j.actamat.2003.12.024.

Pileni M. P., L. Motte and C. Petit 1992, Synthesis of Cadmium sulphide in situ in Reverse Micelles: Influence of the preparation modes on size polydispersity and photochemical reactions, *Chemistry of Material*, 4, 338.

Pillai V., P. Kumar, M. J. Hou, P. Ayyub, and D. O. Shah 1995, Preparation of Nanoparticles of Silver-Halides, Superconductors and Magnetic-Materials Using Water-in-Oil Microemulsions as Nano-Reactors, *Adv. in Colloid and Interface Sci.*, 55, 241-69.

Postma H. W. C., T. Teepen, Z. Yao, M. Grifoni and C. Dekker 2001, Carbon nanotube single-electron transistors at room temperature, *Science*, 293 76-79.

Prakash A., A. P. Bapat and M. R. Zachariah 2003, A simple numerical algorithm and software for solution of nucleation, surface growth, and coagulation problems, *Aerosol Sci. Tech.* 37, 892-98.

Pratsinis S. E. 1997, Precision synthesis of nanostructured particles, in Proc. of the Joint NSF-NIST Conference on Nanoparticles: Synthesis, Processing into *Functional Nanostructures and Characterization* (May 12-13, Arlington, VA).

Pratsinis S. E. 1998, Flame aerosol synthesis of ceramic powders. *Prog. Energ. Combust.* 24,197-219.

Preghenella M., A. Pegoretti and C. Migliaresi 2005, Thermo-mechanical characterization of fumed silica-epoxy nanocomposites, *Polym.*, 46, 12065-72.

Rabe T. and R. Wäsche 1995, Sintering behaviour of nanocrystalline titanium nitride powders, *Nanostruct. Mater.*, 6, 357-60.

Ramamoorthy R., R. N. Viswanath and S. Ramasamy 1995, Synthesis and Study of Nanostructured Yttria Stabilized Zirconia, *Nanostruct. Mater.*, 6, 337-40.

Randolph A. D. and M. A. Larson 1988, Theory of Particulate Processes, Academic Press, New York, pp. 36-37.

Rao N. P. and P. H. McMurry 1989, Nucleation and Growth of Aerosol in Chemically Reacting Systems, *Aerosol Sci. Tech.*, 11, 120-132.

Raut J. S., R. B. Bhagat and K. A. Fichthorn 1998, Sintering of aluminum nanoparticles: A molecular dynamics study, *Nanostructured Materials*, 10, 837-51.

Rueckes T., K. Kim, E. Joselevich, G. Y. Tseng, C. L. Cheung, C. M. Lieber 2000, Carbon nanotube-based nonvolatile random access memory for molecular computing, *Science*, 289, 94-7.

Ruh R., K. S. Mazdiyasni, P. G. Valentine and H. O. Bielstein 1984, Phase Relations in the System $Zr0_2$–Y_20_3 at Low Y_20_3 Contents, *J. Am. Ceram. Soc.*, 67, C-190 -92.

Ryu T. 2008, Chemical Vapor Synthesis (CVS) and Characterization of Tungsten Carbide-Cobalt Nanocomposite Powder, Ph.D. Dissertation, University of Utah, Salt Lake City, Utah.

Ryu T., H. Y. Sohn, K. S. Hwang and Z. Z. Fang 2008a, Tungsten Carbide Nanopowder by Plasma-Assisted Chemical Vapor Synthesis from WCl_6-CH_4-H_2 Mixtures, *J. Mater. Sci.*, 43, 5185–92. http://dx.doi.org/10.1007/ s10853-008-2741-8

Ryu T., H. Y. Sohn, K. S. Hwang, Z. Z. Fang and Y.-U. Kim 2008b, "Plasma Synthesis of Tungsten Carbide Nanopowder from Tungsten Hexachloride," High Temp. Materials and Processes, 27, 91-96.

Ryu T., H. Y. Sohn, G. Han, Y.-U. Kim, K. S. Hwang, M. Mena and Z. Z. Fang 2008c, Nanograined WC-Co Composite Powder by Chemical Vapor Synthesis, *Metall. Mater. Trans. B.* 39B, 1-6. http://springerlink.metapress.com/content/f62766m80166mw23/ fulltext.pdf. http://dx.doi.org/10.1007/s11663-007-9115-5.

Ryu T., H. Y. Sohn, K. S. Hwang and Z. Z. Fang 2008d, Chemical Vapor Synthesis and Characterization of Nanosized WC-Co Composite Powder and Post-Treatment, *Ind. Eng. Chem. Research*, 47, 9384-88.

Ryu T., H. Y. Sohn, K. S. Hwang and Z. Z. Fang 2009a, Chemical Vapor Synthesis (CVS) of Tungsten Nanopowder in a Thermal Plasma Reactor, *Int. J. Refractory Metals and Hard Materials*, 27, 149-54. http://dx.doi.org/10.1016/j.ijrmhm.2008.06.002

Ryu T., H. Y. Sohn, K. S. Hwang and Z. Z. Fang 2009b, Plasma Synthesis of Tungsten Carbide Nanopowder from Ammonium Paratungstate, *J. Am. Ceram. Soc.*, 92, 655–60.

Ryu T., H. Y. Sohn, K. S. Hwang and Z. Z. Fang 2009c, Plasma synthesis of tungsten carbide and cobalt nanocomposite powder, *J. Alloys Compounds*, 481, 274–77. http://dx.doi.org/10.1016/j.jallcom.2009.03.134

Ryu T., K. S. Hwang, Y. J. Choi and H. Y. Sohn 2009d, The Sintering Behavior of Nanosized Tungsten Powder Prepared by a Plasma Process, *Int. J. Refractory Metals and Hard Materials*, 27, 701–04. http://dx.doi.org/10.1016/j.ijrmhm.2008.11.004

Ryu T., H. Y. Sohn, Y.-U. Kim and M. Olivas-Martinez 2010a, Plasma synthesis of nanosized W-Co composite powder followed by carburization with a methane-hydrogen mixture, *J. Nanoparticle Research*, 12, 2851-57. http://www.springerlink.com/openurl.asp?genre=articleandid=doi:10.1007/s11051-010-9870-0.

Ryu T., Y. J. Choi, S. H. Hwang, H. Y. Sohn and I. Kim 2010b, Synthesis of Yttria-Stabilized Zirconia Nanopowders by a Thermal Plasma Process, *J. Am. Ceram. Soc.*, 93, 3130–35. http://onlinelibrary.wiley.com/doi/ 10.1111/j.1551-2916.2010.03843.x/pdf.

Ryu T., M. Olivas-Martinez, H. Y. Sohn, Z. Z. Fang and T. A. Ring 2010c, Computational Fluid Dynamic Simulation of Chemical Vapor Synthesis of WC Nanopowder from Tungsten Hexachloride, *Chem. Eng. Sci.*, 65, 1773–80.

Saeki Y., R. Matsuzaki, H. Nishihara and N. Aoyama 1978, Preparation of cobalt powder by hydrogen reduction of cobalt dichloride, *Denki Kagaku*, 46, No. 12, 643-47.

Sahm T., L. Madler, A. Gurlo, N. Barsan, S. E. Pratsinis and U. Weimar 2004, Flame spray synthesis of tin dioxide nanoparticles for gas sensing, *Sensor. Actuat. B-Chem.*, 98, 148-53.

Sandi G. 2004, Hydrogen storage and its limitations, *Interface,* 13, 40-44.

Sandrock G., J. Reilly, J. Graetz, W.-M. Zhou, J. Johnson and J. Wegrzyn 2006, Alkali metal hydride doping of α-AlH_3 for enhanced H_2 desorption kinetics, *J. Alloys Compd.*, 421, 185-89.

Sara R. W. 1965, Phase Equilibria in the System Tungsten—Carbon, *J. Am. Ceram. Soc.*, 48, 251–257.

Schefflan R., D. Kalyon and S. Kovenklioglu 2001, Modeling of aluminum nanoparticle formation, *Proceedings - Electrochem. Soc.* 2001-13, p. 437-47.

Schlapbach L. and A. Züttel 2001, Hydrogen-storage materials for mobile applications, *Nature* 414, 353-58.

Schmid H. K. 1987, Quantitative Analysis of Polymorphic Mixes of Zirconia by X-Ray Diffraction, *J. Am. Ceram. Soc.*, 70, 367-76.

Schwade B. and P. Roth 2003, Simulation of nano-particle formation in a wall-heated aerosol reactor including coalescence, *J. Aerosol Sci.,* 34, 339-57.

Scott H. G. 1975, Phase Relationships in the Zirconia Yttria System, *J. Mater. Sci.*, 10, 1527-35.

Seinfeld J. H. and S. N. Pandis 2006, Atmospheric Chemistry and Physics. From Air Pollution to Climate Change, 2nd ed., Wiley, Hoboken, NJ, 1225 pp.

Selcuk C. and J. V. Wood 2005, Reactive sintering of porous tungsten: A cost effective sustainable technique for the manufacturing of high current density cathodes to be used in flashlamps, *J. Mater. Process. Technol.*, 170, 471-76.

Setyawan H. and M. Yuwana 2008, Modeling and simulation of aluminum nanoparticle synthesis by the evaporation-condensation process using the nodal method, *Chemical Product and Process Modeling*, Vol. 3, Iss.1.

Shao H., Y. Wang, H. Xu and X. Li 2004, Hydrogen storage properties of magnesium ultrafine particles prepared by hydrogen plasma-metal reaction, *Mater. Sci. Eng. B*, 110, 221-26.

Sheu T.-S., T.-Y. Tien, and I.-W. Chen 1992, Cubic-to-Tetragonal (t') Transformation in Zirconia-Containing Systems, *J. Am. Ceram. Soc.,* 75, 1108-16.

Shi X. L., G. Q. Shao, X. L. Duan, Z. Xiong and H. Yang 2006, Characterizations of WC–10Co nanocomposite powders and subsequently sinterhip sintered cemented carbide, *Materials Characterization*, 57, 358-70. doi:10.1016/j.matchar.2006.02.013.

Siegel R. W. 1994, Nanophase materials, in Ecyclopedia of applied physics, vol. 11, G. L. Trigg, ed., VCH, Weinheim, pp. 1-27.

Singh Gaur R. P. 2006, Modern hydrometallurgical production methods for tungsten, JOM, 58 (9), 45-9.

Sohn H. Y. 1998, Ti and TiAl powders by the flash reduction of chloride vapors, JOM, 50 (9), 50-51.

Sohn H. Y. 2010, Scale-Up Process Simulation, Nanopowder Production from Waste Silicon Sludge, Final Project Report Submitted to KIGAM, University of Utah, Salt Lake City, UT, March 30.

Sohn H. Y. and S. PalDey 1998a, Synthesis of Ultrafine Particles of Intermetallic Compounds by the Vapor–Phase Magnesium Reduction of Chloride Mixtures: Part I. Titanium Aluminides, *Metall. Mater. Trans. B,* 29B, 457–64.

Sohn H. Y. and S. PalDey 1998b, Synthesis of Ultrafine Particles of Intermetallic Compounds by the Vapor–Phase Magnesium Reduction of Chloride Mixtures: Part II. Nickel Aluminides, *Metall. Mater. Trans. B,* 29B, 465–69.

Sohn H. Y. and S. PalDey 1998c, Synthesis of Ultrafine Particles and Thin Films of Ni_4Mo by the Vapor–Phase Hydrogen Coreduction of the Constituent Metal Chlorides, *Mater. Sci. Eng. A*, 247, 165–72.

Sohn H. Y. and S. PalDey 1998d, Synthesis of Ultrafine Nickel Aluminide Particles by the Hydrogen Reduction of Vapor–Phase Mixtures of $NiCl_2$ and $AlCl_3$, *J. Mater. Res.*, 13, 3060-69.

Sohn H. Y. and T. Ryu 2010, Chemical Vapor Synthesis (CVS) of Inorganic Nanopowders, Chapter 5, Nanopowders and Nanocoatings: Production, Properties and Applications, ed. by V. F. Cotler, Nova Science Publishers, Inc., Hauppauge, NY, pp. 147-78,. https://www.novapublishers.com/ catalog/product_info.php?products_id=10391.

Sohn H. Y., Z. Zhang, S. Deevi and S. PalDey 2004, Chemical Vapor Synthesis of Ultrafine Fe-Co Powder, *High Temperature Materials and Processes*, 23, 329-333.

Sohn H. Y., T. Ryu, J. W. Choi, K. S. Hwang, G. Han, Y. J. Choi and Z. Z. Fang 2007, The Chemical Vapor Synthesis of Inorganic Nanopowders, *JOM*, 59 (12), 44-49.

Sohn H.Y., S. Perez-Fontes and J.W. Choi 2010, Computational Fluid Dynamic Modeling of a Chemical Vapor Synthesis Process for Aluminum Nanopowder as a Hydrogen Storage Precursor, *Chem. Eng. J.,* 156, 215-25.

Sohn H. Y., M. Olivas-Martinez and S. E. Perez-Fontes 2011, Mathematical Modeling of Nanopowder Production by Vapor-Phase Processes," Chapter 3, Mathematical Modelling, ed. by C. R. Brennan, Nova Science Publishers, Inc., Hauppauge, NY 11788.

Stenger F., S. Mende, J. Schwedes and W. Peukert 2005, Nanomilling in Stirred Media Mills, *Chem. Eng. Sci.*, 60, 4557-65.

Strobel R., F. Krumeich, W. J. Stark, S. E. Pratsinis and A. Baiker 2004, Flame spray synthesis of Pd/Al_2O_3 catalysts and their behavior in enantioselective hydrogenation, *J. Catal.*, 222, 307-14.

Sunstrom, J. E., IV, W. R. Moser, and B. Marshik-Guerts 1996, General route to nanocrystalline oxides by hydrodynamic cavitation, *Chem. Mater.*, 8, 2061-67.

Swihart M. T. 2003, Vapor-phase synthesis of nanoparticles, *Curr. Opin. Colloid Interface Sci.*, 8, 127-33.

Szekely, J., J. W. Evans and H. Y. Sohn, Gas-Solid Reactions, Academic Press, New York, 1976, pp 70-73.

Tanev P. T. and T. J. Pinnavaia 1996, Mesoporous silica molecular sieves prepared by ionic and neutral surfactant templating: A comparison of physical properties, *Chem. Mater.* 8, 2068-79.

Tang X., R. Haubner, B. Lux and B. Kieffer 1995, Preparation of Ultrafine CVD WC Powders Deposited from WCl_6 Gas Mixtures, *Journal De Physique Iv - J PHYS IV*, 5, C5-1013- 20. DOI: 10.1051/jphyscol:19955119.

Tani E., M. Yoshimura, and S. Somiya 1983, Formation of Ultrafine Tetragonal Zirconia Powders Under Hydrothermal Conditions, *J. Am. Ceram. Soc.*, 66, 11-14.

Tani T., L. Madler and S. E. Pratsinis 2002, Homogeneous ZnO nanoparticles by flame spray pyrolysis, *J. Nanopart. Res.*, 4, 337-43.

Tans S. J., A. R. M. Verschueren and C. Dekker 1998, Room-temperature transistor based on a single carbon nanotube, *Nature*, 393, 49-52.

Tartaj P., O. Bomati-Miguel, A. F. Rebolledo and T. Valdes-Solis 2007, Controlled Release of Precipitating Agents through Solvothermal Destabilization of Microemulsions: One-Pot Synthesis of Monoclinic Zirconia Nanostructures, *J. Mater. Chem.*, 17, 1958-63.

Thornton P. H., R. G. Davies and T. L. Johnston 1970, *Metall. Trans.*, 1, 207-18.

Tong L. and R. G. Reddy 2005, Synthesis of titanium carbide nano-powders by thermal plasma, *Scripta Mater.*, 52, 1253-58.

Tsukada T., S. Venigalla, A. A. Morrone and J. H. Adair 1999, Low-Temperature Hydrothermal Synthesis of Yttrium-Doped Zirconia Powders, *J. Am. Ceram. Soc.*, 82, 1169-74.

Upadhaya G. S. 2002, Cemented tungsten carbide. Noyes Publications, New York.

Van Put J. W. 1995, Crystallisation and processing of ammonium paratungstate (APT), *Int. J. Refract. Met. Hard Mater.*, 13, 61-76.

Vasylkiv O., Y. Sakka and H. Borodinska 2001, Nonisothermal Synthesis of Yttria-Stabilized Zirconia Nanopowder through Oxalate Processing: II, Morphology Manipulation, *J. Am. Ceram. Soc.*, 84, 2484-88.

Veprek S. and A. S. Argon 2001, Mechanical properties of superhard nanocomposites, *Surface and Coatings Technology*, 146-147, 175-182.

Vollath D. 2008, Nanomaterials: An Introduction to Synthesis, *Properties and Application*, Wiley-VCH.

Wagner C. N. J., E. Yang and M. S. Boldrick 1996, The structure of nanocrystalline Fe, W and NiAl powders prepared by high-energy ball-milling, *Nanostruct. Mater.*, 7, 1-2, 1-11.

Wahlberg S., I. Grenthe and M. Muhammed 1997, Nanostructured hard material composites by molecular engineering 1. Synthesis from soluble tungstate salts, *Nanostruct. Mater.*, 9, 105-08. doi:10.1016/S0965-9773(97)00029-9

Wang C.-C. and J. Y. Ying 1999, Sol–Gel Synthesis and Hydrothermal Processing of Anatase and Rutile Titania Nanocrystals, *Chem. Mater.*, 11, 3113–20.

Wang X., Z. Z. Fang and H. Y. Sohn 2008a, Grain Growth during the Early Stage of Sintering of Nanosized WC-Co Powder, *Int. J. Refractory Metals and Hard Materials*, 26, 232-41.

Wang W. N., A. Purwanto, I. W. Lenggoro, K. Okuyama, H. Chang and H. D. Jang 2008b, Investigation on the correlations between droplet and particle size distribution in ultrasonic spray pyrolysis, *Ind. Eng. Chem. Res.*, 47, 1650-59.

Wegner K. and S. E. Pratsinis 2003, Scale-up of nanoparticle synthesis in diffusion flame reactors, *Chem. Eng. Sci.,* 58, 4581-89.

Wei Q., K. T. Ramesh, B. E. Schuster, L. J. Kecskes and R. J. Dowding 2006, Nanoengineering opens a new era for tungsten as well*, JOM*, 58 (9), 40-44.

Welham N. J. 1999, Room temperature reduction of scheelite ($CaWO_4$), *J. Mater. Res.*,14, 619-27.

Westbrook J.H. 1967, Intermetallic Compounds, Wiley, New York.

Winter C. J. and J. Nitsch 1988, Hydrogen as an energy carrier: technologies, systems, economy, Springer-Verlag: Berlin, New York, pp. 16-28.

Won C. W., B. S. Chun and H. Y. Sohn 1993, Preparation of Ultrafine Tungsten Carbide Powder by CVD Method from WCl_6–C_2H_2–H_2 Mixtures," *J. Materials Research*, 8, 2702–08. http://journals.cambridge.org/action/displayAbstract?fromPage=onlineandaid=8169220andfulltextType=RAandfileId=S0884291400071879.

Wu X. Y., W. Zhang, W. Wang, F. Yang, J. Y. Min, B. Q. Wang and J. D. Guo 2004, Ultrafine WC-10Co cemented carbides fabricated by electric-discharge compaction, *J. Mater. Res.* 19, 2240-44. http://journals.cambridge.org/action/displayAbstract?fromPage=onlineandaid=8030252andfulltextType=RAandfileId=S0884291400089779.

Xie H., J. Wang, T. Xi, Y. Liu, F. Ai, and Q. Wu 2002a, Thermal conductivity enhancement of suspensions containing nanosized alumina particles, *J. Appl. Phys.*, 91, 4568-72.

Xie H., J. Wang, T. Xi, Y. Liu 2002b, Thermal conductivity of suspensions containing nanosized SiC particles, *International Journal of Thermophysics*, 23, 571-580.

Xin, X., T. Nishimura, N. Hirosaki, R.-J. Xie and N. Yamamoto 2005, Fabrication of silicon nitride nano-ceramics by high-energy mechanical milling and spark plasma sintering, *Key Eng. Mater.*, 287, 166-70.

Yamaguchi Y. and H. Inui 1993, TiAl compounds for structural application, in Structural Intermetallics, R. Darolia et al. eds., TMS, Warrendale, PA, pp. 127-42.

Yashima M., K. Ohtake, M. Kakihana, H. Arashi and M. Yoshimura 1996, Determination of Tetragonal-Cubic Phase Boundary of $Zr_{1-x}R_xO_{2-x/2}$ (R=Nd, Sm, Y, Er and Yb) by Raman Scattering, *J. Phys. Chem. Solids*, 57, 17-24.

Yih S. W. H. and C. T. Wang 1979, Tungsten – Sources, Metallurgy, Properties, and Applications, Plenum, New York, 500 p.

Zachariah M. R. 1994, Flame processing, in-situ characterization, and atomistic modeling of nanoparticles in the reacting flow group at NIST, in Proc. of the Joint NSF-NIST Conf. on Ultrafine Particle Engineering (May 25-27, Arlington, VA).

Zaluska A., L. Zaluski, and J. O. Ström-Olsen 1999, Nanocrystalline magnesium for hydrogen storage, *J. Alloys Compd.*, 288, 217-25.

Zaluska A., L. Zaluski and J. O. Ström-Olsen 2001, Structure, catalysis and atomic reactions on the nano-scale: a systematic approach to metal hydrides for hydrogen storage, *Appl. Phys. A-Mater. Sci. Process*. 72, 157-165.

Zawrah M. F. 2007, Synthesis and characterization of WC–Co nanocomposites by novel chemical method, *Ceram. Int.*, 33, 155. doi:10.1016/j.ceramint.2005.09.010.

Zhao G. Y., V. V. S. Revankar and V. Hlavacek 1990, Preparation of tungsten and tungsten carbide submicron powders in a chlorine-hydrogen flame by the chemical vapor phase reaction, *J. Less-Common Metals*, 163, 269-80.

Zhu Y. T. and A. Manthiram 1996, Influence of Processing Parameters on the Formation of WC-Co Nanocomposite Powder using Polymer as Carbon Source, *Composites,* 27B, 407-13.

Zucca A., D. L. Marchisio, A. A. Barresi, and R. O. Fox 2005, Implementation of the population balance equation in CFD codes for modelling soot formation in turbulent flames, *Chem. Eng. Sci.*, 61, 87-95.

Züttel A. 2003, Materials for hydrogen storage, *Mater. Today*, 6, 24-33.

Züttel A. 2004, Hydrogen storage methods, *Naturwissenschaften,* 91, 157-72.

INDEX

U

V

W

X

Y

Z